# **MATHS** SQUARED

# MATHS SQUARED

## 100 CONCEPTS YOU SHOULD KNOW

MARIANNE FREIBERGER
AND RACHEL THOMAS

APPLE PRESS

First published in the UK in 2016 by
Apple Press
74–77 White Lion Street
London N1 9PF
United Kingdom

www.quartoknows.com

ISBN 978-1-84543-645-2

QUMMASQ

This book was conceived, designed and produced by
Quantum Books Limited
74–77 White Lion Street
London N1 9PF
United Kingdom

**Publisher:** Kerry Enzor
**Editorial and Design:** Pikaia Imaging
**Editor:** Anna Southgate
**Design:** Dave Jones
**Illustration:** Tim Brown
**Production Manager:** Zarni Win

Printed in China by Toppan Leefung Printing Limited

9 8 7 6 5 4 3 2

# Contents

# Introduction

People have engaged in some form of mathematics since the dawn of civilization – from dividing the spoils of a hunt to counting their children or constructing shelters in which to live. Like other intelligent animals, we are all born with an innate appreciation of numbers and shapes that is essential for our survival.

Over the millennia mathematics has grown into much more than just a simple tool. Today, mathematics is the language of the sciences. It powers our digital world and allows us to create breathtakingly realistic computer games and movie images. It takes us into space and enables us to develop sophisticated medical equipment. Whenever we need to assess something critically – be it the efficacy of a medical drug or the effect of a political policy – we turn to maths. Just why maths should be so effective in describing and understanding the world in which we live is a mystery. The fact is, that whenever we try to describe or investigate patterns and forms, be they visual, physical or mental, we soon apply mathematical ideas, often without even noticing.

But that's only half the story. In its own right, mathematics is a language of profound beauty. Like a piece of music, it winds around the rhythms of logic to weave its own hypnotic structures. Many mathematicians delight in

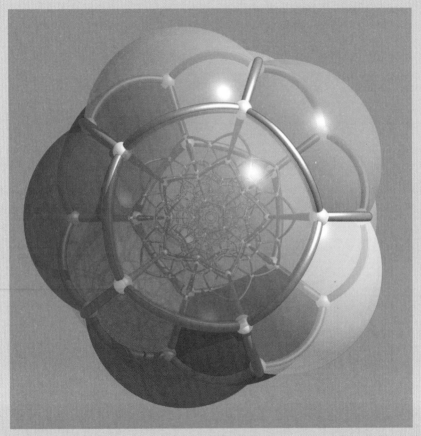

A thing of beauty: the hecatonicosachoron is a four-dimensional analogue of the dodecahedron. Since it is four-dimensional, we can't actually visualize it. This image shows a projection of it into 3D space.

this beauty alone, paying little heed to the applications their discoveries might have. Contrary to popular belief mathematics is a dynamic, ever-changing subject in which there is still much to discover.

In this book we introduce you to 100 concepts we think you would love to know – some because they are useful and others because they are interesting, beautiful or just plain curious. Our path takes you from the basic objects of maths – numbers and shapes – to treacherous geometries, the higher reaches of logic and the expanses of infinity. A degustation menu, if you like, of the very best ingredients and concoctions mathematics has to offer.

Each of the ten chapters is an introduction to a mathematical concept broken down into a further ten brief, but tasty, bites. The first topic is something easy to digest – something you're familiar with, such as simple triangles. And the last topic of each chapter is something to savour: an important result or event that illustrates the boundaries of mathematical knowledge. This includes important unsolved problems – some of which have taunted mathematicians for centuries; important proofs, such as Grigory Perelman's proof of Poincaré's Conjecture in 2003; and mathematical theories, such as Einstein's general theory of relativity.

Rest assured that each and every bite can be digested in just a few minutes. It is our aim to encourage people to appreciate the subject we are lucky enough to enjoy every day, and we hope these tasty morsels whet your appetite to seek out and explore more mathematics. Bon appétit!

Among the subjects covered in this book are the following (as illustrated from top left to bottom right): wallpaper patterns; Peano's axioms; cardinality; hyperbolas; hyperbolic geometry; the butterfly effect; randomness; axiomatic systems; and π (*pi*).

# NUMBERS

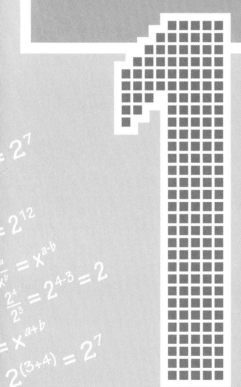

$(2^3)4$

$x^{-a} = \dfrac{1}{x^a}$

$2^{-3} = \dfrac{1}{2^3} =$

$= 2^7$

$= 2^{12}$

$\dfrac{x^a}{x^b} = x^{a-b}$

$\dfrac{2^4}{2^3} = 2^{4-3} = 2$

$x^a x^a$

$2^3 \times 2^4 = 2$

$= x^{a+b}$

$2^{(3+4)} = 2^7$

$(x^a)^b =$

$)^b = x^{ab}$

$2^{12}$

$(2^3)4 = 2^{3 \times}$

$$x^b = x^{a+b}$$

$$= 2^{(3+4)} = 2^7$$

$$b = x^{ab}$$

$$2^{3 \times 4} = 2^{12}$$

$$\frac{x^a}{x^b} = x^{a-b}$$

$$\frac{2^4}{2^3} = 2^{4-3} = 2$$

$$a+b$$

$$4) = 2^7$$

$$2^{12}$$

**W**hen we think about mathematics, numbers are the first things that spring to mind. Most of us meet the counting numbers one, two, three . . . in the first few years of our lives. Our first brush with maths comes when we learn how to add, subtract, multiply and divide these numbers.

In this chapter we take a trip down the number line. We find out how the number zero, while taken for granted today, is a relatively recent invention. We learn that our way of writing numbers – so familiar we hardly ever pay it any attention – is in fact incredibly clever. And we find out how powers give us the power to write, and calculate with, very large and very small numbers.

This chapter also considers a very special class of counting numbers: the primes, which are only divisible by themselves and the number 1.

*Continues overleaf*

In a sense, the primes encode the DNA of all other numbers. It's not surprising, then, that mathematicians love them. They are always on the hunt for larger and larger primes nobody has discovered before, and spend a large amount of time trying to understand how they fit in among all the other numbers. Some of the hardest open problems in maths are related to the prime numbers – we will meet two of them in this chapter.

This chapter doesn't stop at the counting numbers, however. There are also fractions, negative numbers, irrational numbers, complex numbers and more. We meet each of them in turn as we journey through the topics covered, revealing how each conceals its own hidden patterns and structures.

# Contents

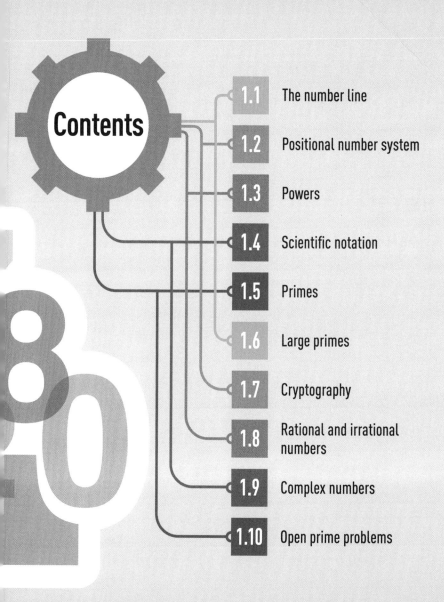

# 1.1 The number line

**The natural numbers have been around ever since people started counting, but zero is a relatively new addition.**

The counting numbers – 1, 2, 3, 4, etc. – come to small children as naturally as they did to early humans, which is why they are called the **natural numbers**. There are infinitely many natural numbers, because you can always add one more. It's useful to think of them as forming a number line – an infinitely long ruler stretching out beyond the horizon.

But what about zero? This figure is a relatively recent addition to mathematics. Although people first started writing down numbers thousands of years ago, it seems that it wasn't until the seventh century that mathematicians in India began to treat zero as a number in its own right. Like the other number symbols we use, the symbol for zero – 0 – originated in India.

Today, we're used to treating zero as an ordinary number. Just like any other number, 0 can be the result of a calculation. If you have £100 in your bank account and withdraw £100, then your balance is 100 – 100 = 0.

If you withdraw £120, you enter the world of negative numbers. Joining those negative numbers (as well as 0) to the number line, gives you a ruler that's infinite in both directions. Doing arithmetic becomes an exercise in moving up and down on this infinite line.

**Some cultures only have words for the first few natural numbers, everything else is called 'many'.**

## Working with negative numbers

When adding:

$$a + (-b) = a - b$$

which means that

$$4 + (-2) = 4 - 2$$

When subtracting:

$$a - (-b) = a + b$$

which means that

$$4 - (-2) = 4 + 2$$

When multiplying:

$$a \times (-b) = (-a) \times b = -(a \times b)$$

which means that

$$4 \times (-2) = (-4) \times 2 = -(4 \times 2)$$

When multiplying negative numbers:

$$(-a) \times (-b) = a \times b$$

which means that

$$(-4) \times (-2) = 4 \times 2$$

The Ishango bone is the earliest evidence of people using tally marks to count. It's around 20,000 years old.

Everyone can perform simple arithmetic when working with positive numbers. Problems start when it comes to working with negative numbers. Life gets easier if you remember a few simple rules.

# 1.2 Positional number system

**Why the way we write numbers is ingenious.**

We have become so used to the way we write the natural numbers that we hardly separate a number from the symbol we use to write it. So it's easy to miss just how clever our number system actually is.

Take the number 423. The symbol 4 does not stand for the number four – rather, it stands for four hundred. Similarly, the symbol 2 stands for twenty. Only the symbol 3 stands for a genuine three. Writing 423 is shorthand for:

(4 × 100) + (2 × 10) + (3 × 1).

The meaning of a symbol depends on its position in the number line. So, reading from the right, the first digit in a number counts how many units (lots of one) there are. The second digit counts lots of ten. The third digit counts lots of one hundred and so on. Each time you move a digit to the left, you multiply the size of the lots you are counting by ten. This idea is clever, because it allows us to write large numbers in an economical way without having to invent new symbols.

Our positional number system is called the **decimal system**, because it is based on the number ten. But it could be based on any other number. For example, digital information is based on the number two and uses only 0s and 1s.

**The Babylonians invented the positional system, but theirs was based on the number 60 rather than the number 10.**

## Babylonian numerals

Above, are some of the numerals used by a people loosely
referred to as the Babylonians. Unlike ours, their system was
based on the number 60.

# 1.3 Powers

**Mathematics is a very efficient language. Working with powers not only makes the maths shorter, it makes calculations easier.**

We can write $2 \times 2$ as $2^2$,

$2 \times 2 \times 2$ as $2^3$, and

$2 \times 2 \times 2 \times 2 \times 2 \times 2 \times 2 \times 2 \times 2 \times 2 \times 2 \times 2 \times 2 \times 2 \times 2 \times 2 \times 2 \times 2 \times 2 \times 2 \times 2 \times 2 \times 2 \times 2 \times 2 \times 2 \times 2 \times 2 \times 2 \times 2 \times 2 \times 2 \times 2 \times 2 \times 2 \times 2 \times 2 \times 2 \times 2 \times 2 \times 2 \times 2 \times 2 \times 2 \times 2 \times 2 \times 2 \times 2 \times 2 \times 2$ as $2^{50}$.

We say we have raised the number 2 to the **power** of 50. Not only is this more efficient than writing the multiplication out in full, but it is also shorter than writing out the actual value of the resulting number:

$2^{50} = 1,125,899,906,842,624.$

Working with powers can make long calculations easier. Calculating the answer to 8,388,608 x 134,217,728 would be impossible in your head and time-consuming using pencil and paper. But calculating the answer to $2^{23} \times 2^{27}$ is simple, thanks to the rules of powers. If you multiply a number raised to one power by that same number raised to some other power, you simply add the powers together:

$2^{23} \times 2^{27} = 2^{23 + 27} = 2^{50}.$

Powers also simplify division: $2^a/2^b = 2^{a-b}$
They also simplify calculating with powers: $(2^a)^b = 2^{a \times b}$.
You can even raise a number to a negative power:
$2^{-a} = 2^0 \times 2^{-a} = 1/2^a.$

Try calculating $1,024^5$. Once you realize that $1,024 = 2^{10}$, things get much easier: $(2^{10})^5 = 2^{10 \times 5} = 2^{50}.$

## Calculating with powers

$$x^a \times x^b = x^{a+b}$$

which means that

$$2^3 \times 2^4 = 2^{(3+4)} = 2^7$$

$$\left(x^a\right)^b = x^{ab}$$

which means that

$$\left(2^3\right)^4 = 2^{3 \times 4} = 2^{12}$$

$$x^{-a} = \frac{1}{x^a}$$

which means that

$$2^{-3} = \frac{1}{2^3} = \frac{1}{8}$$

$$\frac{x^a}{x^b} = x^{a-b}$$

which means that

$$\frac{2^4}{2^3} = 2^{4-3} = 2$$

Here are the rules for working with powers. From top to bottom:
multiplication, raising to another power and division.

# 1.4 Scientific notation

**Raising one number to the power of another quickly calculates very large sums and allows you to express huge (or tiny) numbers using just a few digits.**

Google, the search engine, named itself (bar a misspelling) after a number. In 1929, a **googol** was defined by the American mathematician Edward Kasner as 1 followed by 100 zeros. But rather than writing it out in full – 101 digits in all – you can easily express this number as $10^{100}$ (see Topic 1.3).

Every time you multiply a number by ten, the resulting number is the original with an extra 0 at the end:

$1 \times 10 = 10$,
$1 \times 10^2 = 1 \times 100 = 100$.

So, if you want to write a very large number, you can use this **scientific notation** as shorthand:

$1 \times 10^n = 1$ followed by $n$ zeros.

For example, the speed of light is approximately 300,000,000 metres per second, which we can write using scientific notation as $3 \times 10^8$ m per second. This is much more succinct.

You can also use the same notation to express tiny numbers, by using negative exponents (see opposite). You can write 1/10 as $10^{-1}$, 1/100 as $10^{-2}$ and so on.

**On a calculator $3 \times 10^8$ might appear as 3 e 8 or 3 EX 8.**

## Tiny particles

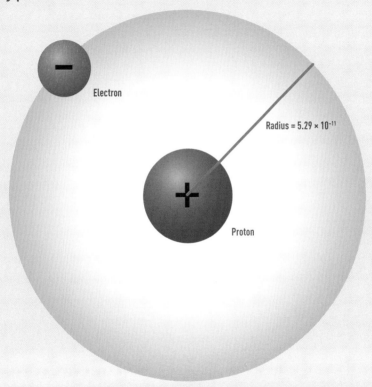

Electron

Radius = $5.29 \times 10^{-11}$

Proton

To write a very tiny number – such as the radius of a hydrogen atom (above) – you can simply write $5.29 \times 10^{-11}$ metres, rather than 0.0000000000529 metres.

# 1.5 Primes

**Primes are the DNA of the natural numbers.**

Some numbers are easily divisible. The number 4, for example, is equal to 2 × 2. Twelve is equal to 2 × 6 or 3 × 4. Not all numbers are so amenable, however.

Try writing the number 3 as a product of two other whole numbers, and the only possibility is 1 × 3. The same goes for 5 (1 × 5) and for 7 (1 × 7). These numbers are examples of **primes**: numbers that can only be divided by the number one and themselves. You can see all the primes less than 100 in the table opposite.

The primes are the mathematical equivalent of atoms: any other whole numbers are a combination of primes. For example:

24 = 2 × 2 × 2 × 3.

Every factor here is a prime number and cannot be divided further. The four prime factors – three 2s and a 3 – are, in a sense, the DNA that uniquely identifies the number 24.

All numbers can be written as a product of primes in a similar way. This result is known as the fundamental theorem of arithmetic and was first proved in ca. 300 BC by the ancient Greek mathematician Euclid of Alexandria. Euclid also showed that there are infinitely many prime numbers – but will we ever find them all? (See Topic 1.6)

**Every prime number is either one more, or one less, than a multiple of six.**

## Prime numbers

| 1 | (2) | (3) | 4 | (5) | 6 | (7) | 8 | 9 | 10 |
|---|---|---|---|---|---|---|---|---|---|
| (11) | 12 | (13) | 14 | 15 | 16 | (17) | 18 | (19) | 20 |
| 21 | 22 | (23) | 24 | 25 | 26 | 27 | 28 | (29) | 30 |
| (31) | 32 | 33 | 34 | 35 | 36 | (37) | 38 | 39 | 40 |
| (41) | 42 | (43) | 44 | 45 | 46 | (47) | 48 | 49 | 50 |
| 51 | 52 | (53) | 54 | 55 | 56 | 57 | 58 | (59) | 60 |
| (61) | 62 | 63 | 64 | 65 | 66 | (67) | 68 | 69 | 70 |
| (71) | 72 | (73) | 74 | 75 | 76 | 77 | 78 | (79) | 80 |
| 81 | 82 | (83) | 84 | 85 | 86 | 87 | 88 | (89) | 90 |
| 91 | 92 | 93 | 94 | 95 | 96 | (97) | 98 | 99 | 100 |

This table of the numbers from 1 to 100 shows the primes and the lowest factors of non-prime numbers.

| Prime | |
|---|---|
| Multiples of 2 | Multiples of 3 |
| Multiples of 5 | Multiples of 7 |

# 1.6 Large primes

**Hunting for ever larger prime numbers has become a popular mathematical hobby.**

We may know for certain that there are infinitely many prime numbers, but unfortunately there is no easy way to spot them. This is in stark contrast to, say, the even numbers. There also are infinitely many of those, but they are easily recognized by the fact that they end in either 0, 2, 4, 6 or 8.

No such trick exists for prime numbers: checking whether a given number is prime can take a huge amount of computing power. This is why the discovery of a hitherto unknown prime always makes a splash in the maths world.

Special candidates for prime numbers are the **Mersenne numbers**. These are numbers that can be written as $2^n - 1$, for some natural number $n$. Examples are:

$3 = 2^2 - 1$ and $7 = 2^3 - 1$.

Mathematical methods for checking whether a Mersenne number is also a prime number are faster than those for checking other numbers, which is why prime hunters tend to focus on them.

Indeed, the last few primes to be discovered were all Mersenne numbers. As of August 2015, the largest known prime number is $2^{57,885,161} - 1$. It isn't written out in full here, because it has over 17 million digits.

**Join the Great Internet Mersenne Prime Search (www.mersenne.org) to help hunt for large primes.**

# The largest known primes

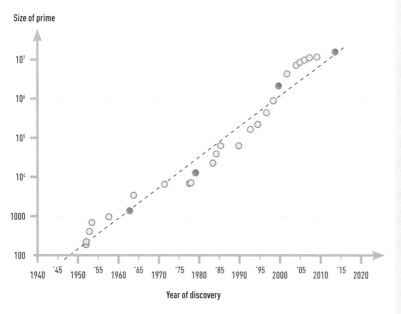

Size of prime

Year of discovery

- 1961: Discovery of the first titanic prime (more than 1,000 digits). It has 1,332 digits.
- 1979: Discovery of the first gigantic prime (more than 10,000 digits). It has 13,395 digits.
- 1999: Discovery of the first mega prime (more than 1,000,000 digits). It has 2,098,960 digits.
- 2013: Discovery of the largest known prime, with a whopping 17,425,170 digits.

This graph shows how the number of digits in the largest known prime has grown over time.

# 1.7 Cryptography

We use cryptography whenever we buy something online, log into a secure website or send secure files.

The Internet relies on a type of cryptography called the RSA **public key cryptosystem**, named for Ronald Rivest, Adi Shamir and Leonard Adleman, who invented it in 1977.

This system allows anyone wanting to receive secure messages (such as an online bank) to advertise a public key. Think of this public key as an open padlock, waiting to be snapped shut on a box containing any message you want to send to the bank. The message is kept nice and secure, and can only be opened with the private key to the padlock, which the bank has never had to share.

This system wouldn't work if anyone could quickly calculate what the private key was from the public key. But the maths behind the RSA system ensures that the only way you can do this, is if you know the factors of a large number, $N$, that makes up the public key.

While multiplying two numbers together is relatively easy, finding the factors of a number can be very difficult. In particular, there are no efficient methods for finding the factors of a number that is the product of very large prime numbers. It is, therefore, a product of large primes that the RSA system uses for the number $N$.

The number $N$ currently used by the RSA system in web browsers and phones has more than 617 digits.

## The RSA System

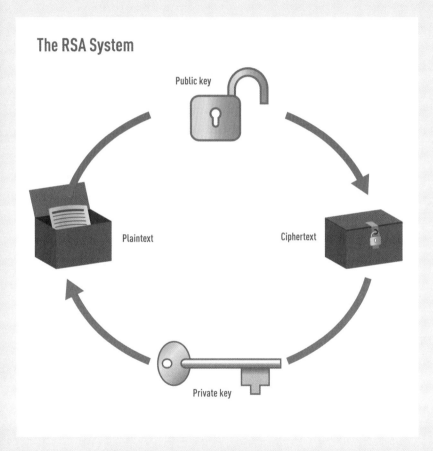

The open padlock and key in the case of the RSA system consist of three numbers. You lock your message up using a public key and the recipient has to use a private key to decrypt the code.

# 1.8 Rational and irrational numbers

**Rational numbers – those that can be represented by a fraction involving whole numbers – have been at the heart of mathematics for several millennia.**

The ancient Greeks, in particular the Pythagoreans, loved the rationals. They believed that all numbers were **rational numbers** (whole numbers were simply fractions with 1 as the lower number: 1 = ½, 2 = ²⁄₁, 3 = ³⁄₁, . . . ).

It seemed to the Greeks that rational numbers might explain the whole universe. An example of this comes from music: play two notes on a stringed instrument, where the length of the string playing one note can be written as a simple fraction of the length of the string playing the other note, and those notes will sound harmonious together. Taking the most basic interval in music – the octave, (the first two notes of the tune 'Somewhere Over the Rainbow') – it is possible to see that it is created by one string being half the length of the other.

Imagine the horror of the Greeks, therefore, when they discovered that not all numbers could be written as fractions of whole numbers. Numbers that aren't rational are **irrational**. One of the first irrational numbers to be discovered was √2 (the square root of 2): the length of a diagonal of a square whose sides have a length of 1. Legend has it that when Greek mathematician Hippasus of Metapontum discovered this number was irrational, he was condemned to death at sea by his fellow mathematicians.

**Together, rational and irrational numbers make up the real numbers.**

## Decimal expansion

■ Rational numbers    ■ Irrational numbers

$$\frac{1}{4} = 0.25$$

$$\frac{1}{3} = 0.3333333333333\ldots$$

$$\frac{1}{7} = 0.142857142857142857\ldots$$

$$\sqrt{2} = 1.414213562373095\ldots$$

$$\pi = 3.141592653589793\ldots$$

$$e = 2.718281828459045$$

The decimal expansion reveals more about these numbers. The decimal expansion of rational numbers either results in a finite number of digits or ends with a repeating pattern. In contrast, the decimal expansion of irrational numbers results in digits that never end and never repeat.

# 1.9  Complex numbers

**Complex numbers were born from the impossible, but have proved incredibly useful.**

Is there a square root of –1? The answer seems to be: no there isn't. Whether that number is positive or negative, multiplying it by itself would always result in a positive answer, which –1 isn't.

In the 16th century, however, mathematicians decided to pretend that there is a square root of –1 (now called an **imaginary number** and denoted by the symbol $i$). The reason for its invention lies in the fact that finding a solution to an equation sometimes involves taking the square root of a negative number. If you pretend that this strange number exists and carry on with your calculations, the result may be a real number, which is also the solution to your equation.

By using the number $i$ you can build **complex numbers**. These have the form $a + ib$, where $a$ and $b$ are real numbers. Examples are $1 + 2i$ or $3 + 5i$. There are rules for doing arithmetic using complex numbers (see opposite), and so they form a perfectly coherent number system.

Outside of maths, complex numbers are useful when something is best described using pairs of numbers. For example, in electronics, current and voltage are more complicated than simple numbers, so their features are better captured when represented as complex numbers.

**The square root of a number $n$ is the number which, when multiplied by itself, gives $n$.**

## An Argand diagram

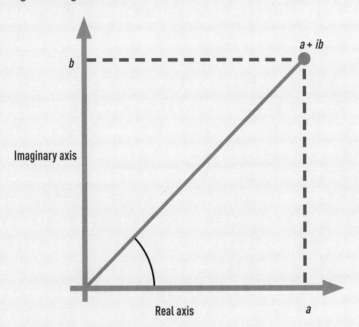

$$(a + ib) + (c + id) = (a + c) + i(b + d)$$
$$(a + ib) \times (c + id) = (ac - bd) + i(bc + ad)$$

Complex numbers have a geometrical interpretation. The complex number $a + ib$ (shown above by the green dot) corresponds to the point with coordinates $(a, b)$ in a Cartesian coordinate system (see Topic 3.2).

# 1.10 Open prime problems

**Many open questions in number theory are simple to state – any one of us could understand them. Yet the answers to these questions have eluded mathematicians for centuries.**

In 1742, Christian Goldbach (1690–1764) wrote a letter to Leonhard Euler (1707–83) containing what became known as *Goldbach's Conjecture*: that every even number greater than 2 can be written as a sum of two prime numbers. For example:

$4 = 2 + 2$
$6 = 3 + 3$
$8 = 5 + 3$.

Every mathematician believes this conjecture to be true, and it has been checked by computers for every even number up to $4 \times 10^{17}$. However, complete proof remains elusive. There was great excitement in 2013 when a young Brazilian mathematician, Harald Helfgott (b. 1977), announced a proof of something called the Weak Goldbach Conjecture: that every odd number greater than 5 is the sum of three prime numbers. The name 'weak' comes from the fact that this results directly from the Goldbach conjecture being true: if we know an even number is a sum of two primes, then we can add the prime number 3 to get all the odd numbers being the sum of three primes.

Mathematicians believe that Helfgott's proof is correct. Yet, despite some impressive mathematics, it doesn't look as if the techniques he used will give us the proof of the full Goldbach conjecture.

**Initially, Euler treated Goldbach's letter with disdain, regarding this still unsolved problem as trivial!**

379

# Twin primes

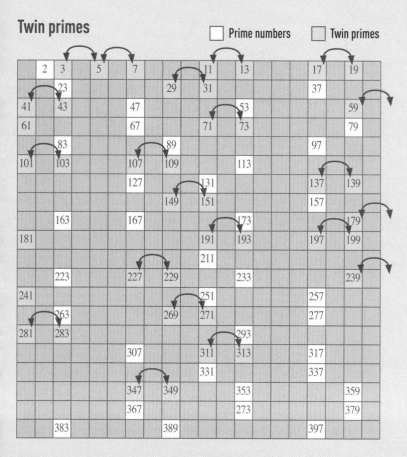

☐ Prime numbers    ☐ Twin primes

Another elusive proof comes from twin primes – primes that differ by
2. Mathematicians believe there are infinitely many twin primes, but
so far no one has been able to prove this.

# SHAPES

**A**t around the same time that children first learn to count, they also start to draw their first geometric shapes: triangles, squares and circles. Unsurprisingly, geometry is one of the first mathematical pursuits humanity engaged in. Not only do those shapes come naturally to us, we also need to understand them in order to live our lives – to measure the fields we cultivate, for example, and to build the houses in which we live. The word *geometry* derives from the Greek words for 'Earth' and 'measurement'. In fact, today we continue to use the basic rules of geometry first devised by the ancient Greek scholars.

In this chapter we explore those ideal shapes beloved by school children and ancient mathematicians alike. The circle gives us one of the most famous numbers in mathematics, as well as the most efficient way to enclose area.

*Continues overleaf*

The humble triangle is the cornerstone of trigonometry, which explores the relationship between a triangle's angles and its sides. And an ingenious connection between circles and triangles gives us some of the most useful functions in mathematics.

Through Euclid's rules for flat geometry, triangles provide an essential part of the definition of space as we know it. Triangles were also the key mathematicians needed to discover a whole new type of geometry. It is here that we start to venture into realms that we find hard to picture in our minds.

As well as looking at new geometries, we will also discover seemingly impossible shapes, we'll find out why coffee cups and doughnuts are mathematically the same, and that all of us already live our lives in higher dimensions. Finally, we'll discover how all of these ideas are vital ingredients in a recent proof of a famous centuries-old problem.

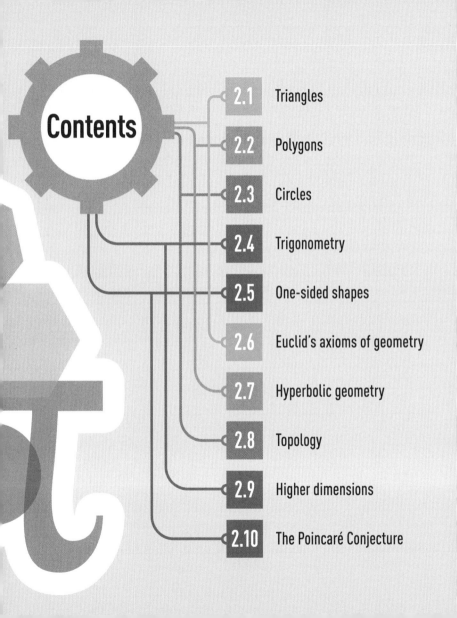

# Contents

# 2.1 Triangles

The triangle is the first shape we learn about in school. It is deceptively simple, yet hides a wealth of mathematical and physical power.

Most of us, when asked to draw a triangle, automatically draw something close to an **equilateral triangle**: one with all sides of equal length and all angles of equal size.

The triangles we see in the world around us – either in the bracing of a crane or bridge or in the frame of a swing in the playground – are almost always symmetrical, with one half identical to the other. These are called **isosceles triangles**: they have two sides of the same length and opposing angles of equal size. In a **scalene triangle**, all three sides are of different lengths, and each of the three angles is of a different size.

The angles of a triangle drawn on a flat piece of paper always add up to 180 degrees. There is a relationship between the size of an angle and the length of the opposite side of the triangle: the bigger the angle, the longer the opposite side. A triangle in which each of the angles is less than 90 degrees is called an **acute triangle**, while a triangle in which one angle is greater than 90 degrees is **obtuse**.

The special case of a triangle in which one angle is equal to 90 degrees – a **right triangle** – is the basis of one of the most famous theorems in the history of mathematics: the Pythagorean theorem (see Topic 3.10).

> The triangle is the strongest straight-sided shape: you can deform a square with hinged corners, but a hinged triangle will remain rigid.

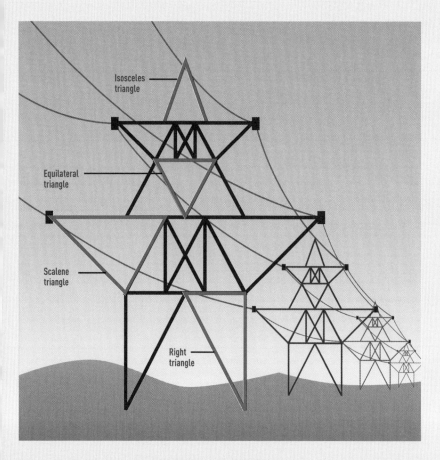

Pythagoras' theorem: for a right triangle, the square of the length of
the longest side (opposite the right angle, called the hypotenuse) is
equal to the sum of the squares of the lengths of the other two sides.

# 2.2 Polygons

**From the humble triangle to the perfect circle in infinitely many, regular steps.**

A step up from the triangle is the rectangle. In fact, more precisely, a step up from the triangle is the **quadrilateral**: a closed shape bounded by four straight sides.

The rectangle belongs to a special class of quadrilaterals: those whose corner angles all measure 90 degrees. If each of the four sides also has the same length, then you have a square. Similar closed shapes exist with five sides (pentagons), six sides (hexagons), seven sides (heptagons), eight sides (octagons), nine sides (nonagons) and so on.

Once the number of sides grows large, mathematicians drop the Greek names and instead describe a shape using a strange, but practical, combination of a number and the suffix '-gon' (from the Greek word *gonia* meaning 'corner' or 'angle'). A 96-sided shape is called a 96-gon, for example, and a 200-sided shape would be a 200-gon. Collectively, these shapes are known as **polygons** – which loosely translates as 'shapes with many corners'.

The equilateral triangle and the square are both **regular polygons**, whose sides all have the same length and the same angles between them. There are also regular pentagons, hexagons, heptagons and so on, however far you'd like to go.

**The ancient Greeks used polygons to approximate the circumference and area of a circle.**

## Polygons and circles

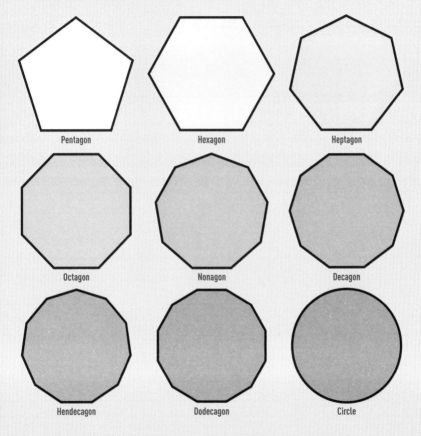

Pentagon

Hexagon

Heptagon

Octagon

Nonagon

Decagon

Hendecagon

Dodecagon

Circle

Although the sides of many-sided polygons are still straight lines, they become so short in relation to a shape's circumference that, together, they approximate a circle's smooth roundness.

# 2.3 Circles

**A circle is the collection of points that lie at a fixed distance (the radius) from a given point (the centre).**

Circles are beautifully regular. Every straight line that passes through the centre of a circle is a line of symmetry – it is possible to reflect the circle in this line without changing its appearance. Similarly, a circle can rotate around its centre through any angle while maintaining exactly the same shape. This makes the circle the most symmetrical of all shapes.

Another beautiful property of a circle is that it maximizes the area it encloses. If you were given a piece of rope and asked to lay it out in a closed shape so that the area inside is as large as possible, then you should choose the shape of a circle: any deformation of the circle would result in an inward bulge, so reducing the internal area.

The circle also gives us one of the most famous numbers in mathematics. Take any circle of any size, and you will find that, if you divide its circumference by its diameter, the result is equal to 3.14159 . . . , a constant that is ubiquitously known as $\pi$ (the Greek letter *pi*; see opposite). The area of the circle is always equal to $\pi r^2$, where $r$ is the **radius** of the circle (measured as a straight line from the circle's centre to any point on its circumference).

**As of 2015, humanity has managed to calculate the first 13.3 trillion decimal places of $\pi$.**

3.14159265358979323846264338327950288419716939937510582097494459230781640628620899862803482534211706798214808651328230664709384460955058223172535940812848111745028410270193852110555964462294895493038196442881097566593344612847564823378678316527120190914564856692346034861045432664821339360726024914127372458700660631558817488152092096282925409171536436789259036001133053054882046652138414695194151160943305727036575959195309218611738193261179310518548074462379962749567351885752724891227938183011949129833673362440656643086021394946395224737190702179860943702770539217176293176752384674818467669405132000568127145263560827785771342757789609173637178721468440901224953430146549585371050792279689258923542019956112129021960864034418159813629774771309960518707211349999998372978049951059731732816096318595024459 . . .

The number π is irrational, which means that its decimal expansion is infinitely long and doesn't settle into a repeating pattern.

# 2.4 Trigonometry

What do astronomy, navigation, geography and architecture have in common? For a large part of their history they relied on trigonometry, the 'measuring of triangles', for success.

Essentially, **trigonometry** explores the relationships between the sizes of the angles and the lengths of the sides in a triangle. We'll call one of the two smaller angles in a right triangle θ (it could be either one). Then we can define the basic trigonometric relationships using the lengths of the three sides of the triangle:

sin θ = length of the opposite side/length of the hypotenuse.

cos θ = length of the adjacent side/length of the hypotenuse.

Note that the **hypotenuse** is the longest side – the one opposite the right angle – and the *adjacent side* is the one running from the angle θ to the right angle.

If you keep the same length hypotenuse, but change the size of the angle θ, then the length of the opposite side will vary with θ and so will the value of sine θ. Similarly, the length of the adjacent side will vary and so will the value of cosine θ.

The resulting changing values of the trigonometric functions define smooth repeating waves. These sine and cosine waves lie at the heart of analysing any periodic fluctuations, from sound waves to seismology.

**For any angle θ,**
$$cos^2(θ) + sin^2(θ) = 1.$$

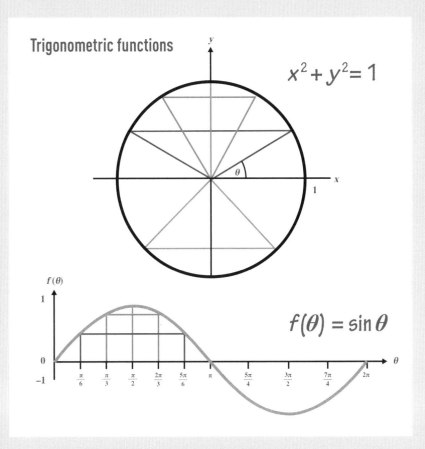

**Trigonometric functions**

$x^2 + y^2 = 1$

$f(\theta) = \sin \theta$

This graph shows how the function sin $\theta$ (solid blue wavy line) can be drawn using the angles inside triangles. Here, we have the graph of the function $y = \sin \theta$ for $\theta$ in the interval $(0, 2\pi)$.

# 2.5 One-sided shapes

Take a strip of paper, give it a twist and then join the ends together. You have created an important mathematical object: a Möbius strip.

A regular strip of paper has two sides – the front and the back. But how many sides does a **Möbius strip** have? Once you have made a Möbius strip, draw a line down its centre without lifting the pencil. Keep going until you reach the point at which you started.

Had you joined the ends of your paper strip without giving it a twist, you'd find that you had only drawn a loop around the inside or around the outside of the strip. With a Möbius strip, however, you can draw a line to connect any two points (even if those points were on opposite sides of the paper strip originally). So a Möbius strip doesn't have an inside and outside: it has just one side.

These kinds of one-sided shapes are referred to as **non-orientable** shapes. Just as it has only one side, a Möbius strip also has only one edge. You can check this by running a felt-tip pen along the edge until you return to where you started. The edge on 'both' sides of the strip will be coloured in one continuous loop.

One-sided shapes are important mathematically. Möbius strips are the fundamental example of non-orientability: they are contained within every possible non-orientable surface.

Conveyor belts and typewriter ribbons have been constructed as Möbius strips, to make better use of the surface area on 'both sides' of the strip.

# The Klein bottle

This image was created by Charles Trevelyan, who rendered the bottle partially transparent so that we can appreciate its form fully.

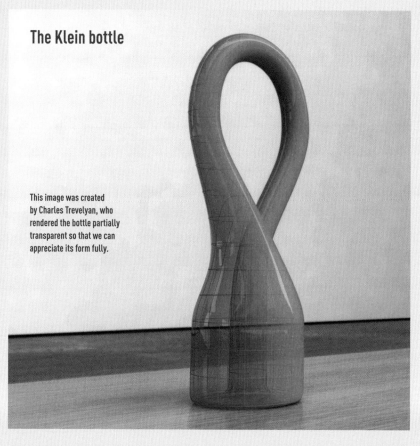

The Klein bottle has only one side and encloses no volume. It can't exist in three-dimensional space, as an extra fourth dimension is needed in order for the neck to pass through the wall of the bottle without intersecting it.

# 2.6 Euclid's axioms of geometry

**Geometry goes back all the way to the ancient Greek mathematician Euclid of Alexandria.**

The ancient Greeks were fond of geometry. Not content simply with drawing shapes, however, they also liked to prove that results, such as the Pythagorean theorem, were definitely true.

The only way to prove something is to derive it logically from basic facts that you know for sure are true. This is why Euclid of Alexandria (ca. 300 BC) devised his five **axioms** of geometry (see opposite):

- **1** Given any two points, you can draw a straight line between them.
- **2** Any line segment can be made as long as you like.
- **3** Given a point $P$ and a line segment $l$ starting at $P$, you can draw a circle centred on $P$ with $l$ as its radius.
- **4** All right angles are equal to each other. (To Euclid a right angle was an angle constructed in a particular way. The axiom asserts that all angles constructed in this way are equal.)
- **5** The fifth axiom is hard to state in simple terms, but equates to the fact that the angles in a triangle add up to 180 degrees.

**Some say that Euclid did not exist, and that *The Elements* was written by several mathematicians.**

Euclid published his axioms in *The Elements*, one of the earliest known rigorous treatments of geometry and other areas of maths. It is one of the most successful books ever – some say only the bible has had more editions.

## Euclid's axioms

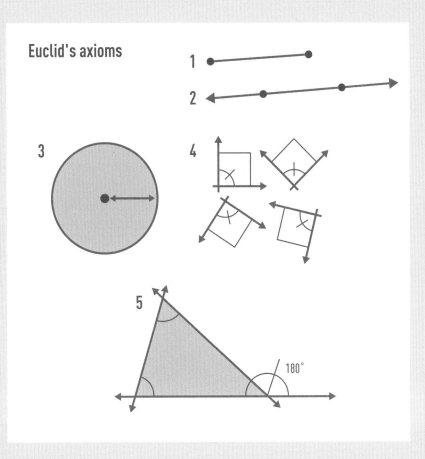

Euclid's five axioms. Using his five axioms, Euclid was able to prove
mathematical results through geometric constructions.

# 2.7 Hyperbolic geometry

**After trying – and failing – to prove that Euclid's fifth axiom always holds, mathematicians came up with a wonderful new type of geometry.**

When the 19th-century mathematician János Bolyai (1802–60) spent time trying to find a proof that Euclid's fifth axiom is always true, he received a letter from his father Farkas Bolyai (1775–1856) beseeching him to give it up.

This axiom (see Topic 2.6) is equivalent to saying that the angles in a triangle always add up to 180 degrees. János' father feared that the work involved, *'may take all your time, and deprive you of your health, peace of mind and happiness in life'*. He may have been right.

The axiom does hold for triangles drawn on a flat plane. A triangle drawn on a sphere, however, bulges outwards, causing the angles to add up to more than 180 degrees. (The sides of the spherical triangle are analogous to straight lines – that is, paths of shortest distance.) Is there a surface on which the angles in a triangle add up to less than 180 degrees? The answer is yes, and an example of this is a surface shaped like a saddle.

In their misguided hunt for a proof of Euclid's fifth axiom, mathematicians eventually developed **hyperbolic geometry**, in which Euclid's fifth axiom does not hold. Triangles in the **hyperbolic plane** always have angles adding up to less than 180 degrees.

**Hyperbolic geometry is an important ingredient in Einstein's special theory of relativity.**

# The hyperbolic plane

This picture shows the whole of the hyperbolic plane. Distortion means that the tiles appear to get smaller towards the edge of the circle. However, in the actual hyperbolic metric, they are all the same size.

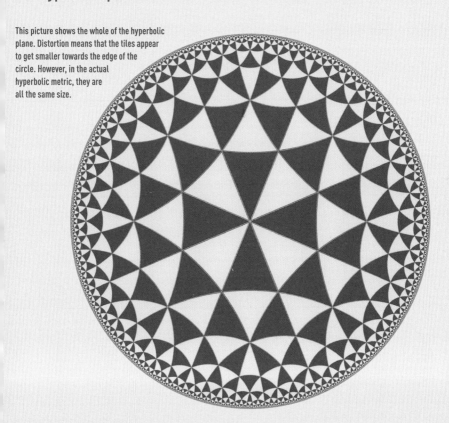

It's impossible to represent the hyperbolic plane on a flat piece of paper without distortion.

# 2.8 Topology

**Geometry is very precise, while its sister subject, topology, is generally more permissive.**

In topology, two shapes are regarded as the same if one can be transformed into the other without cutting, tearing or gluing. A famous example of this concerns a coffee cup and a doughnut: if the coffee cup were made from an elastic material, you could deform it into the shape of a doughnut, turning the hole formed by the handle into the hole of the doughnut.

Allowing shapes to bend, squeeze and stretch in this way can be useful. Take the map of the London Underground. Geographically, the map is woefully inadequate: the distances between stations are distorted and all the lines appear to run straight, be it vertically, horizontally or at 45 degree angles.

Yet, a geographically accurate map would be a tangled mess, with stations jam-packed in the centre. It would also have to be huge to accommodate the outer reaches of the network. The Tube lines are not straight, but weave confusingly around the city and each other.

It was the London Underground employee Harry Beck (1902–74) who realized, in 1933, that the map could be laid out better, focusing simply on the connections between lines. His topological map has become an iconic image, regarded by many as a piece of design genius.

**In topology, even a knobbly potato is a perfect sphere.**

## Seamless transformation

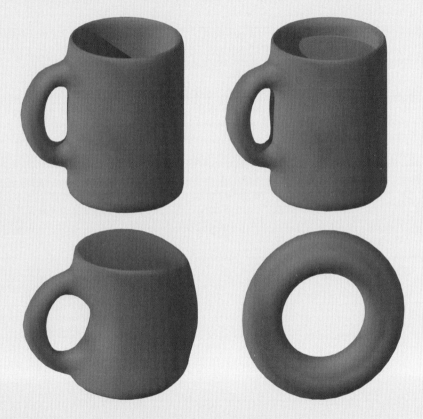

In the above example, the cup and the doughnut are considered the same shape in topological terms, since one can be transformed into the other without making any cuts.

# 2.9 Higher dimensions

**Higher dimensions sound daunting but, in fact, most of us are already quite used to them.**

If you arrange to meet a friend in a high-rise building, you need four pieces of information: the street name, the street number, the floor of the building and the time at which you intend to meet.

Mathematicians call these pieces of information *coordinates*, and these four **coordinates** define a four-dimensional space. A five-dimensional space is defined by a set of five coordinates, six-dimensional spaces by a set of six coordinates and so on. To think in higher dimensions, therefore, mathematicians simply think in terms of more and more coordinates.

Many familiar concepts can be extended to higher dimensional spaces. For example, a circle is the set of points in two dimensions that are all the same distance from the centre. A sphere is the set of points in three dimensions that are all the same distance from the centre, and a hypersphere is the set of points in four or higher dimensions that are all the same distance from the centre (where the idea of distance in four and higher dimensions is simply an extension to the one with which we are familiar). The definition is identical in two-, three-, four- or higher-dimensional spaces. It is simply the number of coordinates needed to describe the locations of the points in each space that changes.

**Instead of having a boundary consisting of six square faces, like the cube, the four-dimensional tesseract is enclosed by eight cubic faces.**

## Towards the fourth dimension

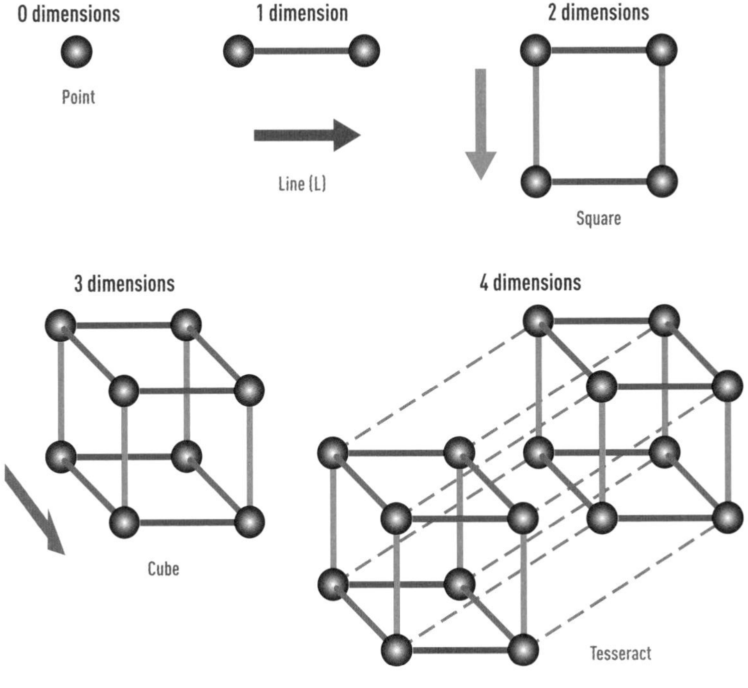

**0 dimensions**

Point

**1 dimension**

Line (L)

**2 dimensions**

Square

**3 dimensions**

Cube

**4 dimensions**

Tesseract

A square is made by starting with a line segment of length L, and extending it by a distance of L to make a square. A cube extends a two-dimensional square into the third dimension, and a tesseract extends a three-dimensional cube into the fourth dimension.

# 2.10 The Poincaré Conjecture

**In topology, a sphere doesn't have to be perfectly round. You can mould it into any shape and as long as you don't puncture it, that shape remains a sphere.**

So what defines a topological sphere, if not its roundness? One answer comes from loops. Any loop drawn on a sphere can, in theory, be shrunk down to a single point. This isn't the case for a doughnut, however: if the loop winds through its hole, there's no way of shrinking it down without cutting the loop. The fact that loops can be shrunk to points without cutting is what uniquely differentiates the topological sphere from other surfaces in the same family.

If the above is true for a two-dimensional sphere existing in a three-dimensional space, is the same true for a three-dimensional sphere existing in a four-dimensional space? While it may not be possible to visualize this sphere, it – and its loop-shrinking property – can be rigorously described mathematically. At the start of the 20th century, the French mathematician Henri Poincaré (1854–1912) asserted that an analogue of the loop-shrinking result also works for a three-dimensional sphere.

However, neither Poincaré, nor countless mathematicians after him, could prove this *Poincaré Conjecture*, which became notorious as a result. The conjecture wasn't solved until nearly a century later, when the Russian mathematician Grigori Perelman (b. 1966) published the proof in a series of online papers. Perelman was duly offered prizes and honours, but refused them all. Not everyone is motivated by fame and fortune.

**In 2000 the Clay Mathematics Institute offered $1 million for a resolution of the Poincaré Conjecture.**

## Shrinking loops

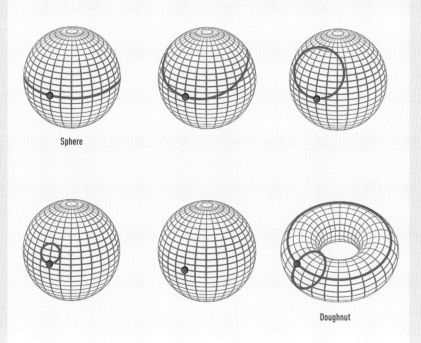

Sphere

Doughnut

The red loop encircling the sphere could be shrunk down to a point.
The loop through the hole of the doughnut can't.

# EQUATIONS

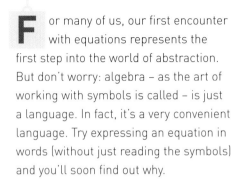

**F** or many of us, our first encounter with equations represents the first step into the world of abstraction. But don't worry: algebra – as the art of working with symbols is called – is just a language. In fact, it's a very convenient language. Try expressing an equation in words (without just reading the symbols) and you'll soon find out why.

In this chapter we will meet the basic components of mathematical equations. We will see how equations are linked to some of the familiar shapes we know from geometry and find out that some equations can be solved and others can't. We'll learn that, at one time, fierce mathematical duels were fought over their solutions. We'll also meet a few special equations and one of the hardest problems in mathematical history: Fermat's last theorem.

*Continues overleaf*

Before we start, however, here's a beautiful example of an equation phrased in words. It's a problem from the 12th century maths book *Lilavati* by the Indian mathematician Baskara II.

*Of a herd of elephants, half together with a third part of itself was roaming in a forest; a sixth part together with a seventh of itself was drinking water from a river; an eighth part together with a ninth of itself was playing with lotuses. The leader of the herd was seen with three females. What was the number of elephants in the herd?*

Translated into an equation and simplified a little, this is simply $0.996x + 4 = x$.

# Contents

# 3.1 Variables and constants

**The power of abstraction captured by a few friendly symbols.**

If you earn £$x$ a month, how much do you earn in a year? The answer is 12$x$. Writing $y$ for your yearly income, gives you: $y = 12x$.

This is an example of an **equation**. The symbol $x$ is called a **variable** because, in theory, it could stand for any number at all. The symbol $y$ is a **dependent variable** because its value depends on $x$. And the number 12 is called a **constant** for the obvious reason that it doesn't change.

Using symbols to represent numbers lands us firmly in the world of algebra. For a more general equation, we could also write a symbol, for example $a$, to represent the constant 12. The equation $y = ax$ would then capture any situation in which a variable $y$ is equal to some fixed number (namely $a$) times another variable $x$, no matter if the $x$ stands for your income, say, or the price of a burger.

Algebra is taken for granted today, but this step into abstraction was a great advance in the development of mathematics. The use of symbols didn't really take hold in maths until the 15th century, however. Before this time, people expressed equations using words, which could be pretty tedious.

**The word 'algebra' comes from the Arabic *al-jabr*, meaning 'restoration'.**

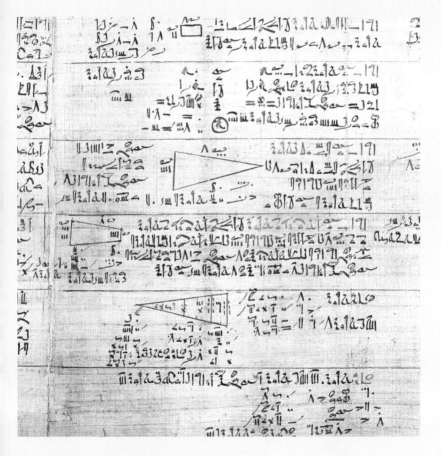

One of the first records of algebraic abstraction exists in the
5.5-m-long (18-foot) Rhind papyrus, published in Egypt
ca. 1650 BC. It is currently held at the British Museum, London.

# 3.2 Cartesian coordinates

We often think of algebra and geometry as different areas of maths, but in fact there is a deep connection between the two fields.

One day, the French mathematician René Descartes (1596–1650) lay in bed, contemplating a fly on the wall (or so the story goes). He wondered how best to describe the location of the fly and, as a result, he came up with what we now call **Cartesian coordinates**.

To specify a point in the plane (for example the fly's location on the wall), first draw two perpendicular axes meeting at a point we'll call $O$. Each point is then determined by two coordinates: the first gives its horizontal distance to $O$, and the second its vertical distance to $O$. The first coordinate is usually called the $x$-coordinate and the second is called the $y$-coordinate.

And the connection to algebra? Suppose you have the equation $y = 2x$. Now look at all the points whose coordinates $(x,y)$ satisfy this equation. In other words, find the points whose coordinates take the form $(x, 2x)$. The point $(0,0)$ qualifies here, as do the points $(1,2)$ and $(2,4)$.

A little more thought will reveal the fact that the points whose coordinates satisfy this equation all lie along a straight line that connects $(0,0)$, $(1,2)$ and $(2,4)$. In fact, the equation exactly determines this line.

The equation $x^2 + y^2 = 2^2$ defines a circle with centre $(0,0)$ and radius 2.

# The Cartesian coordinate system

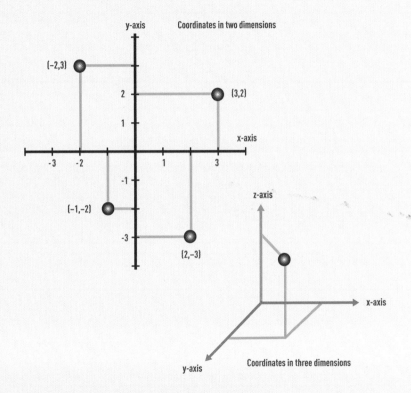

The diagrams above show coordinates in two and three dimensions.
The connection between algebra and geometry allows mathematicians
to solve problems in algebra using geometry and vice versa.

# 3.3 Quadratic equations

**Quadratic equations are those in which the highest power of the variable is 2. They are useful in many contexts.**

The size of a square field whose sides are $x$ m long is $x^2$. In a **quadratic equation**, smaller powers of $x$ can also exist. For example, the size of another field, in which one side is 2 m longer than the other, is expressed as:

$$x(x + 2) = x^2 + 2x.$$

In general, any quadratic equation can be written as:

$$y = ax^2 + bx + c.$$

Where $x$ is the variable that changes; $a$, $b$, and $c$ are the coefficients; and the value of $y$ depends on the value of $x$. A graph of the values of $y$ for different values of $x$ (see Topic 3.2) results in a shape called a **parabola**.

Parabolas have a fascinating property: any line entering a parabola parallel to its centre line reflects off the curve of the parabola and passes through a point on its centre line (see opposite). This point is called the **focus**. It is for this reason that the curve of a satellite dish, if you cut the dish in half, is shaped like a parabola. The dish is angled in such a way that the radio waves it receives enter the dish parallel to the centre line. They then reflect off the surface of the dish and pass through the focus, which is where the signal receiver is located.

**The lightbulb of a car headlight sits at the focus of a parabolic mirror.**

## Parabolic dish

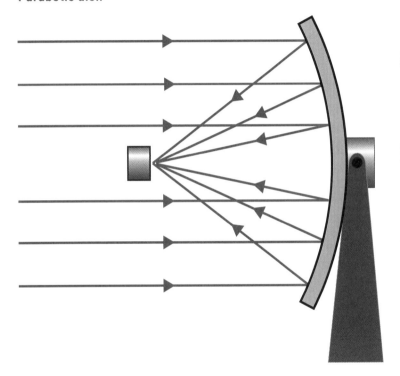

What do the flight of a ball thrown through the air, the graceful arc of water from a fountain and the shape of a satellite dish have in common? They can all be described using a quadratic equation.

# 3.4 Cubic equations

In the 16th century mathematicians challenged each other to duels, their weapons being mathematical techniques for solving equations. A favourite challenge was to solve cubic equations.

**Cubic equations** are those in which the highest power of the variable $x$ is 3. For example: $x^3 + 2x - 33 = 0$, which has the solution $x = 3$.

You might think solving these should be relatively easy, as we all learnt the rule for solving quadratic equations in school (see Topic 3.3). The value of $x$ in a quadratic equation – expressed as $ax^2 + bx + c = 0$ is:

$$x = \frac{-b \pm \sqrt{b^2 - 4ac}}{2a}$$

This rule – the **quadratic formula** – has existed since at least AD 628, but a similar general rule for cubic equations proved elusive. Anyone who found a technique for solving a particular type of cubic equation, therefore, kept it secret. For example, people developed a method to solve depressed cubics, equations of the form $x^3 + bx + c = 0$, but didn't want to share it.

The Italian mathematician Girolamo Cardano (1501–76) put an end to these jealously guarded secrets. He'd learned the method for solving **depressed cubics** independently from two sources, one of whom, Tartaglia, made Cardano swear an oath of secrecy. Cardano published the method in his great work, the *Ars Magna* (*The Great Art*) and used this method in an ingenious way to find a general solution that could solve any cubic equation.

The *Ars Magna* contained the first example of calculating with the square root of a negative number (see Topic 1.9).

## Depressed cubics

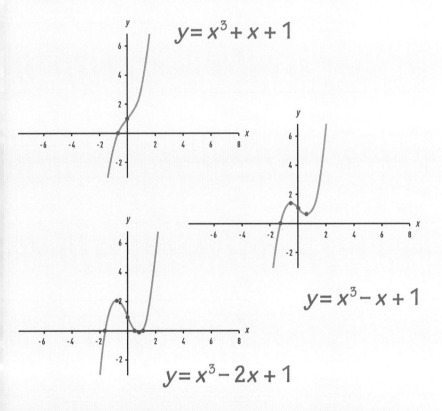

$$y = x^3 + x + 1$$

$$y = x^3 - x + 1$$

$$y = x^3 - 2x + 1$$

These three graphs are examples of depressed cubic equations – that is, those without a squared power. The method for solving depressed cubics had been independently developed by Scipione del Ferro (1465–1526) and Niccolò Fontana (ca. 1500–57).

# 3.5 Quintic equations

**The search for a solution for quintic equations led to the birth of symmetry and featured the work of two tragic heroes.**

What is $x$, if $x^5 = 32$? The answer is $x = 2$. This shows that it's perfectly possible to solve a **quintic equation** – one in which the highest power of $x$ is 5.

The question is whether there is a **general solution**: a formula akin to the quadratic formula that gives you the solution to any quintic equation. Perhaps surprisingly, the answer is: no. This result was proved in 1824 by the 22-year-old Norwegian mathematician Niels Henrik Abel (1802–29). Unfortunately, Abel died in poverty five years later, from tuberculosis.

Not long after, the Frenchman Évariste Galois (1811–32) decided to understand why the quintic equation doesn't admit a general solution. Galois noticed that, with equations, there often seems to be a symmetry at work. A glimpse of this symmetry is revealed by noticing that, if $x$ is a solution to $x^5 = 32$, then $-x$ is a solution to $x^5 = -32$. It's as if $x$ and $-x$ mirror each other.

Galois consequently developed a theory of symmetry (see Topic 7.1), explaining why there isn't a general solution to the quintic equation. Unfortunately, Galois also came to a tragic end: in 1832 he was killed in a duel at just 20 years of age.

**Galois' work laid the foundations of group theory, a fundamental pillar in the world of mathematics.**

## Perfect symmetry

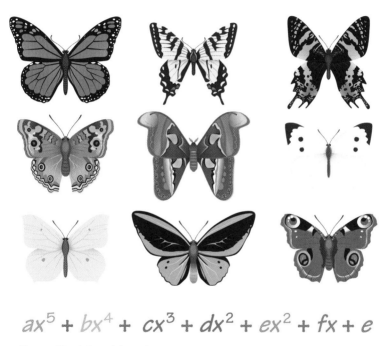

$$ax^5 + bx^4 + cx^3 + dx^2 + ex^2 + fx + e$$

The general formula for a quintic equation.

Galois laid the foundations for group theory – the mathematical study of symmetry. It has many applications in physics, a discipline in which the underlying equations are often assumed to exhibit symmetries.

# 3.6 Polynomials

Expressions made up of multiples of powers of a variable that are added to or subtracted from each other are called polynomials.

Examples of **polynomials** are:

$x^4 + 2x^3 - 3x^2 + 4x + 5$
$2x^{10} - 10x^5 + 7x^3 + 9x^2 + 4x + 17$.

Solving such equations isn't always an easy task (see Topic 3.5), yet polynomials have another beautiful property: other mathematical expressions can be written in terms of them. As an example, consider trigonometry (see Topic 2.4). The cosine of $x$, $\cos(x)$, can be written as:

$$\cos(x) = 1 - \frac{x^2}{2!} + \frac{x^4}{4!} - \frac{x^6}{6!} + \ldots$$

This infinitely long polynomial is called a **power series**, and it has a beautiful pattern to it. A complementary expression works for the sine of $x$, $\sin(x)$.

There are infinite polynomials for many mathematical expressions, and they are useful in many mathematical contexts. For example, if you need to work $\sin(x)$ or $\cos(x)$ for some value of $x$, but your calculator doesn't have the appropriate button, you can approximate the answer by working out the first few terms of the power series.

For a number $n$, the product $n \times (n-1) \times (n-2) \times (n-3) \times \ldots \times 2 \times 1$ is called $n$ factorial and abbreviated to $n!$ (see opposite).

## Polynomial graph

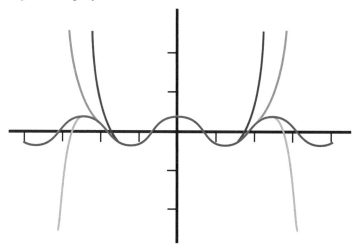

$$f(x) = 1 - \frac{x^2}{2!} + \frac{x^4}{4!} - \frac{x^6}{6!} + \frac{x^8}{8!}$$

$$f(x) = 1 - \frac{x^2}{2!} + \frac{x^4}{4!} - \frac{x^6}{6!} + \frac{x^8}{8!} - \frac{x^{10}}{10!} + \frac{x^{12}}{12!}$$

$$f(x) = 1 - \frac{x^2}{2!} + \frac{x^4}{4!} - \frac{x^6}{6!} + \frac{x^8}{8!} - \frac{x^{10}}{10!} + \frac{x^{12}}{12!} - \frac{x^{14}}{14!}$$

The magenta curve represents cos(x). The other curves represent the approximation to cos(x) using the first few terms of the power series.

# 3.7 Power laws

**Surprisingly many phenomena in the world, whether natural or human-made, are described by one particular class of equations.**

This class consists of equations of the form:

$$y = 1/x$$
$$y = 1/x^2$$
$$y = 1/x^3$$

and so on. Whenever a variable $y$ varies in proportion to $1/x^k$, for some number $k$, we say that it follows a **power law**.

An important class of processes that follow **power laws** involves networks – be they friendship networks, the Internet, the power grid or transportation networks. In such a network each node (for example, a person) is linked to a number of other nodes (their friends). If you count the number $y$ of nodes that are linked to exactly $x$ other nodes, you often find that the relationship between $x$ and $y$ is similar to $y = 1/x^k$, where $k$ is usually a small number, somewhere between 2 and 4.

This universality of power laws may seem surprising, but mathematicians have shown that it may be the result of a simple mechanism they call 'the rich get richer'. Assuming that a network grows by new nodes always choosing to connect with nodes that already have lots of connections, you can show that a relationship of the form $y = 1/x^k$ emerges quite naturally.

**Earthquakes follow a power law, where y is the number of earthquakes of magnitude x.**

# The internet

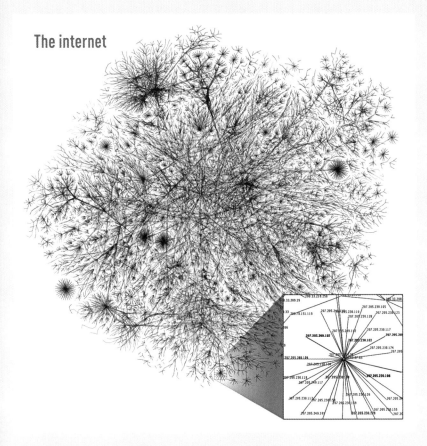

Many networks exhibit similar features, including the power law behaviour for link distributions. This image shows a portion of the internet, based on data from 2005.

# 3.8 Compound interest and *e*

**Debt is no fun, but it's nice to know that one of the most important numbers in mathematics, *e*, is hiding behind those interest calculations.**

Suppose you borrow £100 at an annual interest rate of 100% (admittedly a little unrealistic). If the bank calculates your total amount owed at the end of the first year, you will owe your initial amount, £100, plus the interest amount, £100. But what if the bank decides to calculate the interest quarterly, using one-quarter of the interest rate each time?

After the first three months you would owe:
$100 + \frac{1}{4} \times 100 = 100 \times (1 + \frac{1}{4}) = £125$.

After six months you would owe: $100 \times (1 + \frac{1}{4}) + \frac{1}{4} \times (100 \times (1 + \frac{1}{4})) = 100 \times (1 + \frac{1}{4})^2 = £156.25$.

And after one year: $100 \times (1 + \frac{1}{4})^4 = £244.14$ – more than twice the amount you borrowed.

If the bank **compounds** the interest $n$ times a year using $\frac{1}{n}$th of the rate, the total owed increases by a factor of $(1 + \frac{1}{n})^n$. The more times they compound (that is, the greater value for $n$), the more you owe.

Thankfully there is a limit, $e$. If you calculate this compounding factor for higher and higher values of $n$, it brings you closer and closer to (but never exceeds) a value of:

$e = 2.71828182845904523536028747135266249775724$ $709369995 \dots$

**The number *e* can often be used to rewrite complicated equations describing growth in a much simpler way.**

## Rising interest

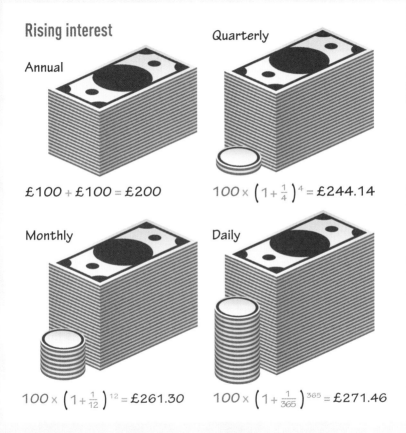

**Annual**

£100 + £100 = £200

**Quarterly**

$100 \times \left(1 + \frac{1}{4}\right)^4 = £244.14$

**Monthly**

$100 \times \left(1 + \frac{1}{12}\right)^{12} = £261.30$

**Daily**

$100 \times \left(1 + \frac{1}{365}\right)^{365} = £271.46$

The amount you owe after a year increases the more times the interest is calculated throughout that period. This growth, or compounding, of the interest is the curse of debt and, equally, the blessing of saving.

# 3.9 Euler's identity

Ask any mathematician what the most beautiful maths equation is, and the answer will likely be Euler's identity.

This is **Euler's identity**:

$e^{i\pi} + 1 = 0$.

Why is it beautiful? One reason is that this equation contains important mathematical numbers: $e$, the number that captures growth; $i$, the square root of -1 and the heart of the complex numbers; $\pi$, from circles and geometry; and 0 and 1, the building blocks of our number system.

Another reason is the equation's simplicity. It is based on a formula that Leonhard Euler (1707–83) developed when he was considering how the number $i$ fitted in with the rest of mathematics:

$e^{i\theta} = sin\ \theta + i\ cos\ \theta$.

It turns out that a complex number describes the position of a point in the plane, and both sides of this equation describe the same complex number (see opposite). We get the point $e^{i\pi}$ if we rotate the angle $\theta$ around to 180 degrees (this is equivalent to $\pi$ in *radians*, another measure of angles), which is the point (–1,0). So from Euler's formula we get:

$e^{i\pi} = -1 + 0$.

You can easily rearrange this to find Euler's identity.

These few symbols elegantly convey a wealth of knowledge that is the culmination of centuries of mathematical work.

## Euler's formula

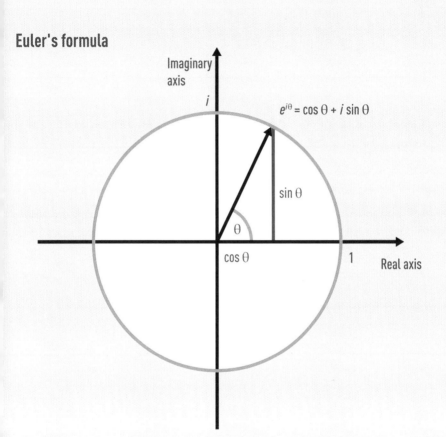

The left-hand side of Euler's formula describes the position of a point in terms of an arrow starting at the origin that has length 1 and makes an angle of Θ with the horizontal axis. The right-hand side describes the point using coordinates (cos Θ, sin Θ).

# 3.10 Fermat's last theorem

**Fermat's last theorem, one of the most celebrated mathematical results of the last century, is based on the humble right triangle.**

As shown in the image opposite, there are many triples of whole numbers that satisfy the Pythagorean theorem. These **Pythagorean triples** led 17th-century French mathematician, Pierre de Fermat (1601–65), to wonder if there were whole number triples for powers higher than 2. That is, whole numbers that satisfy:

$a^3 + b^3 = c^3$ or $a^4 + b^4 = c^4$ and so on.

He began to suspect that, surprisingly, there weren't. This conjecture – that there are no such whole number triples for powers greater than 2 – is now known as *Fermat's last theorem*.

Fermat noted his conjecture in the margin of a book, finishing with the line: *I have discovered a truly marvellous proof of this, which this margin is too narrow to contain.*

This line taunted mathematicians for more than 350 years, until the British mathematician Sir Andrew Wiles surprised the world with a proof in 1993. In fact, he had proved a more general result that, as a consequence, confirmed Fermat's last theorem. After first announcing the result, it took another year of work for him and a colleague to fix a few mistakes. The final proof was over 150 pages long and contained new mathematical techniques that have had a great impact in mathematics.

**Wiles worked in secret on the proof of Fermat's last theorem for seven years.**

There are many examples of whole numbers *a*, *b* and *c* that form the sides of a right triangle. Such sets of three numbers, (*a*, *b*, *c*) are called Pythagorean triples. There are infinitely many such triangles, some examples of which are shown above.

# LIMITS

4

**M**athematicians like to take things to the limit. In life, as well as in mathematics, it's good to know where you're going. In this chapter we'll explore what we intuitively mean by a sequence of steps converging to some final destination.

We first meet this idea in the most famous number sequence in maths which, when followed to infinity, gives us a finite number that appears everywhere from spiral galaxies to the leaves on a plant. This number, called the golden ratio, is also infamous as the most irrational of the irrational numbers – a fact that is revealed by writing it in a special way, as a continued fraction. Writing numbers in this way also reveals special properties of the rational numbers and patterns hidden inside many irrationals too.

*Continues overleaf*

The final destinations of infinite sequences are called limits. The language of limits has allowed mathematicians to describe that great constant in life – change. Calculus is used to describe change with great power and precision. Its discovery was at the heart of one of the bitterest disputes in mathematics between its independent discoverers, Isaac Newton and Gottfried Leibniz.

In this chapter we'll also discover how to cut a cake into infinitely many slices and how careful slicing of time means an underdog (or, in this case, an undertortoise) can avoid losing a race. Perhaps Newton and Leibniz should have just resolved their differences over coffee and cake, or settled it with a race.

# Contents

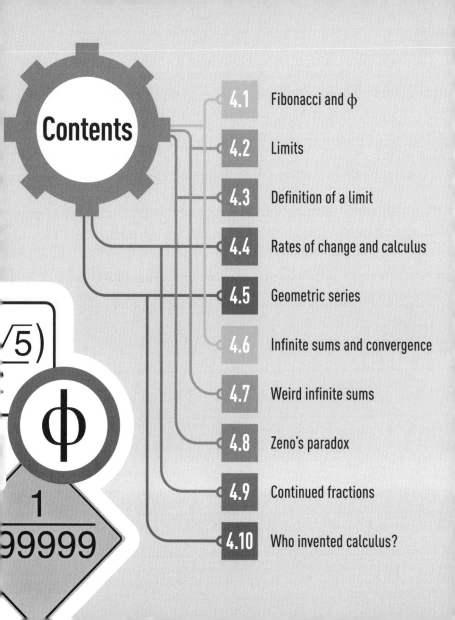

# 4.1 Fibonacci and φ

**A famous number sequence takes maths to the limit.**

In the following number sequence, each number (from 2 onwards) is the sum of the two preceding numbers:

1, 1, 2, 3, 5, 8, 13, 21, 34, 55, 89, 144, . . .

The sequence is named for the Italian mathematician, Fibonacci (ca. 1170–ca. 1250), who wrote about it in 1202. It has a very interesting feature – a new sequence emerges when dividing each term by the one that precedes it:

$\frac{1}{1}$, $\frac{2}{1}$, $\frac{3}{2}$, $\frac{5}{3}$, $\frac{8}{5}$, $\frac{13}{8}$, $\frac{21}{13}$, $\frac{34}{21}$, $\frac{55}{34}$, $\frac{89}{55}$, $\frac{144}{89}$, . . .

When written in decimal (rounded to four places), it looks like this:

1, 2, 1.5, 1.6667, 1.6, 1.6250, 1.6154, 1.6190, 1.6176, 1.6181, 1.6180.

Moving along this new sequence, the numbers always seem to stay in the region of 1.618. In fact, they get arbitrarily close to a very special number called $\phi$ (*phi*):

$\phi$ = 1.618033988 . . . , which can also be written as

$$\phi = \frac{(1+\sqrt{5})}{2}$$

The number $\phi$ crops up in many geometric shapes and, along with the Fibonacci sequence, can also be found in growth patterns, for example of plants. For these reasons it's one of the most famous constants in maths.

**φ is called the golden ratio and has been known since the ancient Greeks. It is also the first example of a limit.**

# The Fibonacci spiral

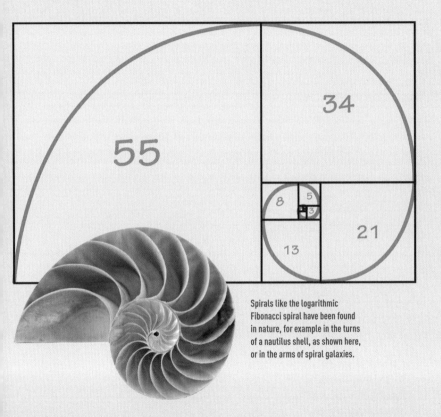

Spirals like the logarithmic
Fibonacci spiral have been found
in nature, for example in the turns
of a nautilus shell, as shown here,
or in the arms of spiral galaxies.

The Fibonacci spiral is made by drawing circular arcs connecting
opposite corners of squares whose side lengths (the numbers in the
squares above) are given by the Fibonacci sequence.

# 4.2 Limits

**Many number sequences approach a limit and can do so in different ways.**

Consider the following number sequence:

$\frac{1}{2}$, $\frac{2}{3}$, $\frac{3}{4}$, $\frac{4}{5}$, $\frac{5}{6}$, $\frac{6}{7}$, . . .

Moving along the sequence, the numbers grow bigger and bigger, yet never exceed 1. This is a curious concept – numbers in a sequence can grow larger and larger, yet never exceed a certain limit. This is because the steps by which the numbers increase become smaller and smaller, and no step is ever large enough to go beyond 1. You can find a similar behaviour with decreasing sequences:

$\frac{1}{2}$, $\frac{1}{3}$, $\frac{1}{4}$, $\frac{1}{5}$, $\frac{1}{6}$, $\frac{1}{7}$, . . .

In this case, the numbers grow smaller and smaller, but never plummet below 0. You can even have oscillating sequences that behave in a similar way, for example:

$\frac{1}{2}$, $-\frac{1}{3}$, $\frac{1}{4}$, $-\frac{1}{5}$, $\frac{1}{6}$, $-\frac{1}{7}$, $\frac{1}{8}$, $-\frac{1}{9}$, . . .

Here, the numbers increase at one step but decrease at the next. They don't jump about randomly, however. Instead, they zero in on 0.

This illustrates one of the most important concepts in mathematics: number sequences can **converge** to a certain **limit**. This idea is central to calculus (see Topic 4.4), the driving force behind a vast range of applications of maths. Without limits, maths would itself be limited.

**Archimedes, famous for shouting 'Eureka' in the bath, was one of the first mathematicians to use limits.**

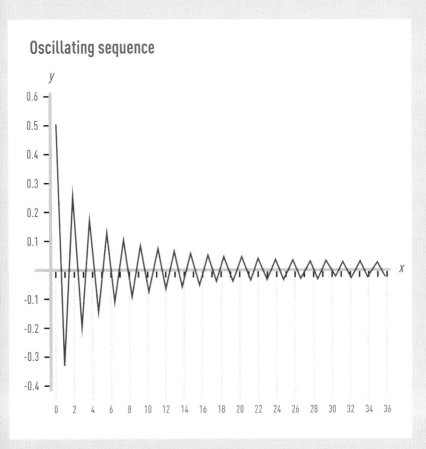

## Oscillating sequence

This graph illustrates the oscillating sequence ½, –⅓, ¼, –⅕, . . .
The points approach 0 from either side of the *x* axis, without ever
getting there.

# 4.3 Definition of a limit

Intuitively, it's clear what is meant by saying that a number sequence converges to a limit. Actually describing this in words, however, is quite a challenge.

Here is the correct definition of a **limit**. Don't be scared, you'll see it's actually very clever:

'A number sequence converges to a limit $x$ if, given *any* positive number $\varepsilon$ (the Greek letter, *epsilon*), which can be as small as you like, you can find a number $N$, which might have to be very large, so that all terms in the sequence that come after the $N$th term lie within a distance of $\varepsilon$ of $x$.'

The idea is this. Whenever you choose a number $\varepsilon$, which could be really small, say, 0.0000001, there will always be a number in the sequence, call it $a$, that's within $\varepsilon$ of $x$. This captures the fact that numbers in the sequence get arbitrarily close to the limit $x$. The number $N$ is there to rule out the possibility that after $a$ the sequence veers away from $x$ again.

Attempts to define a limit in any other way, will soon run into difficulties. It is with this definition that mathematical language really comes into its own.

**Every mathematician knows that $\varepsilon$ is always a very small number.**

## Approaching a limit

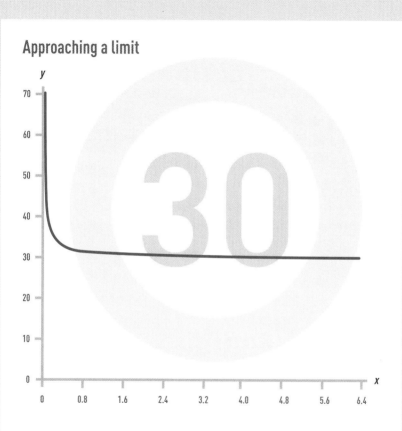

This graph tracks the value of the sequence $30 + 1/x$ for $x = 1, 2, 3, \ldots$ You need to go to the 10000000th term to get within 0.0000001 of the limit 30. All the terms after this point are within this distance of 30.

# 4.4 Rates of change and calculus

It is said that the only constant in life is change. It is, therefore, vital to understand how things change. Calculus offers the perfect way to describe this.

We experience **rates of change** everywhere in life: speed is the rate of change of distance, acceleration describes the rate of change of speed, and power is the rate of change in work – all measured with respect to time.

Measuring these rates of change involves breaking time up into chunks. For example, one chunk could cover the time it takes you to climb a ladder. Then, the rate of change in the distance you covered is simply the length of the ladder, per that unit in time. You could break the time up further, for each step of the ladder. Then the rate of change in the distance you covered is the length between rungs of the ladder, per the new unit in time. This will be constant if you take the same amount of time with each step. However, if you start to tire and climb more slowly, the rate of change in the distance you covered will decrease, even though you move forwards with each step.

**Calculus** is the process of calculating the rate of change of some quantity, allowing the time steps to become smaller and smaller – to become **infinitesimal** – until you can calculate the rate of change at any instant.

We can't do calculus at sharp corners, breaks or jumps – the quantity needs to vary smoothly.

## Playground fun

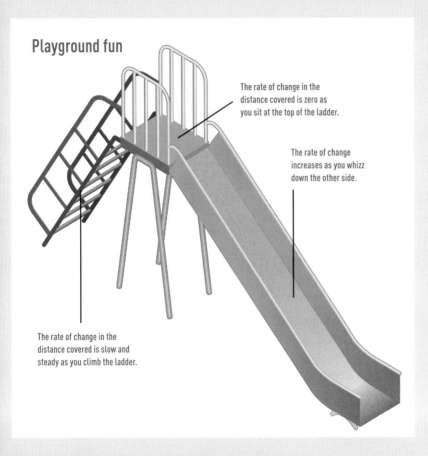

The rate of change in the distance covered is zero as you sit at the top of the ladder.

The rate of change increases as you whizz down the other side.

The rate of change in the distance covered is slow and steady as you climb the ladder.

To understand the joy of calculus, think about the various stages of taking a go on a playground slide. First you climb the ladder. Then, you may pause a moment at the top, before whooshing down the other side.

# 4.5 Geometric series

**How do you divide one cake between infinitely many people? You use the geometric series.**

When you distribute a cake as in the picture opposite each person receives a piece that is half the size of the piece given to the previous person. The first person has half of the cake. The second person has half of a half:

$$\tfrac{1}{2} \times \tfrac{1}{2} = \tfrac{1}{4} = \tfrac{1}{2}^2,$$

and the size of the $n$-th person's piece is $\tfrac{1}{2}^n$.

If we sum the sequence of piece sizes, the result is what's called a **geometric series**:

$$\tfrac{1}{2} + \tfrac{1}{4} + \tfrac{1}{8} + \tfrac{1}{16} + \tfrac{1}{32} + \ldots + \tfrac{1}{2}^n + \ldots,$$

and this must add to 1: the whole cake.

Amazingly, an infinite sum can have a finite result. A geometric series is one in which each term is some constant ratio, $r$, times the previous term. In general, a geometric series has a finite sum if, and only if, the ratio $r$ is strictly less than 1.

**The total cake given away after the first $n$ pieces is $1 - \tfrac{1}{2}n$.**

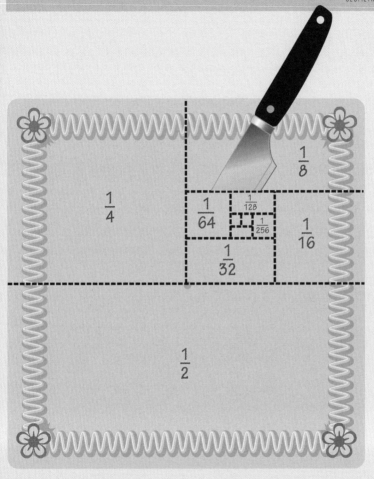

Give the first person half the cake. Give the second person half of the remaining piece. Give the next person half of what's left and so on. After every cut, you'll always have some cake left over.

# 4.6 Infinite sums and convergence

The previous topic suggests the sum of an infinite sum is equal to 1. But what do we mean by an infinite sum having a finite value?

Consider the **partial sums** of an infinite sum – the sums that are created by adding one more term each time. The $n$th partial sum of the geometric series from the previous page is:

$$S_n = \tfrac{1}{2} + \tfrac{1}{4} + \tfrac{1}{8} + \tfrac{1}{16} + \tfrac{1}{32} + \ldots + \tfrac{1}{2}^n.$$

These partial sums form a sequence of numbers $S_1$, $S_2$, $S_3$ and so on. We know what it means for a sequence of numbers to converge (see Topics 4.2 and 4.3). If a sequence of partial sums converges to a limit, then we say the infinite sum **converges**.

Obviously, for an infinite sum to have a finite answer you need the terms to get smaller and smaller, but this isn't always enough. For example, the **harmonic** series:

$$1 + \tfrac{1}{2} + \tfrac{1}{3} + \tfrac{1}{4} + \tfrac{1}{5} + \ldots$$

doesn't converge to a finite limit, even though the individual terms $\tfrac{1}{n}$ are getting closer to zero. This is because the partial sums of the series don't converge. In fact, the partial sums become infinitely large and so we say this sum is **divergent**.

The harmonic series diverges very slowly – you have to add more than $10^{43}$ terms to get a partial sum that is over 100!

# Harmonic series

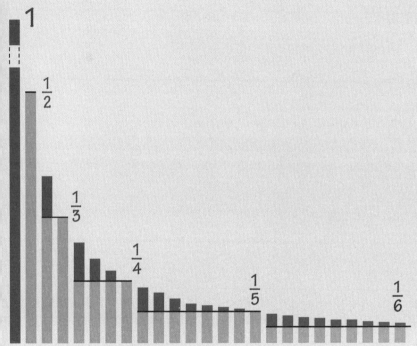

The sum of the harmonic series is divergent.

You can divide the terms of the harmonic series into groups, where each group adds to more than one-half. This can be seen in the image above, where the blue bars in each group add up to exactly ½. Adding infinitely many halves gives you an answer of infinity.

# 4.7 Weird infinite sums

**Things can become very entertaining with infinite sums.**

Consider the following infinite series:

$S = 1 - 1 + 1 - 1 + 1 - 1 + \ldots$

We can group the terms in pairs, as follows:

$S = (1 - 1) + (1 - 1) + (1 - 1) + \ldots$

Calculating each bracket results in: $S = 0 + 0 + 0 + \ldots$

So it seems that $S$ is equal to 0. But what if we write $S$ as:

$S = 1 + (-1 + 1) + (-1 + 1) + (-1 + 1) + \ldots$

This gives $S = 1 + 0 + 0 + 0 + \ldots$, so $S$ should be equal to 1.

Thus, there are good reasons to believe that the sum is equal to both 0 and 1 at the same time. This is down to the fact that the infinite sum doesn't converge in the ordinary sense explored in Topic 4.6. With such **divergent** sums, you can play all sorts of fun games. And things get even weirder when you look at the **alternating harmonic series**:

$1 - \frac{1}{2} + \frac{1}{3} - \frac{1}{4} + \frac{1}{5} - \frac{1}{6} + \ldots$

**When summed in its usual order, the oscillating harmonic series converges to $ln(2) \approx 0.69$.**

Simply by rearranging the order of terms, you can make this sum converge (in the ordinary sense) to any number you like! It goes to show: infinite sums need to be treated with care.

# The Casimir effect

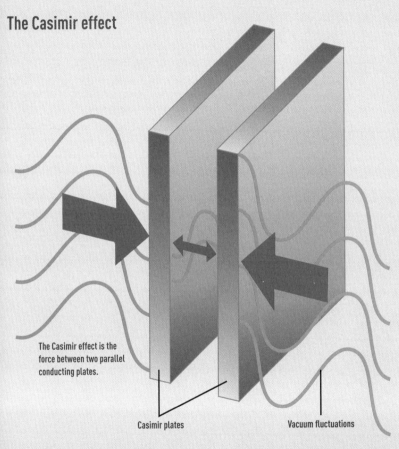

The Casimir effect is the force between two parallel conducting plates.

Casimir plates

Vacuum fluctuations

It is possible to show that $1 + 2 + 3 + 4 + 5 + \ldots = -\frac{1}{12}$. Divergent infinite sums like this one enter into physics. The result that the infinite sum is given by a finite quantity actually predicts a phenomenon called the Casimir effect. Experiments confirm this piece of mathematical trickery.

# 4.8 Zeno's paradox

**The ancient philosopher Zeno proposed a paradox that continues to puzzle us today: how does a slow tortoise manage to outrun the Greek warrior Achilles?**

Suppose the race is set up as in the image opposite. Then, after one minute Achilles reaches the point, $T_0 = 100$ m, where the tortoise started from. At this time the tortoise has already travelled 1 m and so is at the point $T_1 = 101$ m. Achilles will reach $T_1$ after another 0.01 minutes, after which time the tortoise will have moved another 0.01 m and be at the point $T_2 = 101.01$ m and so on.

Each time Achilles reaches the point that the tortoise was last at, the tortoise has sneaked ahead and it seems like Achilles will never catch up. But we know that he should finish the race in 1000/100 = 10 minutes, while the tortoise will cross the finish line later, in (1000−100)/1 = 900 minutes.

Achilles' distance, as described by Zeno, can be written as:

$100 + 1 + 0.01 + 0.001 + 0.0001 + \ldots$

**In effect, Zeno is slowing down time so Achilles never gets to pass the tortoise.**

This is a **geometric series** (see Topic 4.5). As the ratio of this series, $r = 0.01$, is less than 1, we know that this infinite sum will converge. In fact it converges to 101.010101 . . ., a point 1.010101. . . minutes after the start of the race, and the point at which Achilles will overtake the tortoise. Zeno is dividing up time and distance into ever smaller pieces, creating the impression that this point is never reached.

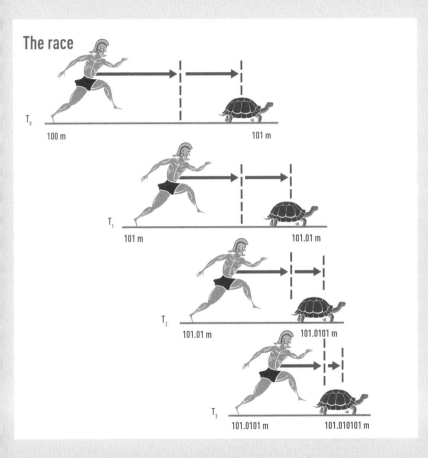

**The race**

$T_0$
100 m                    101 m

$T_1$
101 m                    101.01 m

$T_2$
101.01 m                 101.0101 m

$T_3$
101.0101 m               101.010101 m

Achilles races the tortoise over 1,000 m. He runs at 100 m per minute
and the tortoise travels along at a leisurely 1 m per minute. The
tortoise is so slow that he has a head start of 100 m. Who wins?

# 4.9 Continued fractions

**Every number can be written as a continued fraction – a nested series of fractions. Continued fractions are finite for rational numbers and infinite for irrational numbers.**

While an irrational number can't be written as a simple fraction, it can be approximated by one: for example, $\pi$ (*pi*) is approximately $^{22}/_7$. A fraction $^p/_q$ is a *good* approximation of an irrational number $x$ if there isn't a fraction that is closer to $x$ with a smaller denominator $q$. And a **continued fraction** of a number gives the best approximations, in this sense. You can calculate these by cutting off an infinite continued fraction. For example, the first five good approximations of:

$$\pi = 3 + \cfrac{1}{7 + \cfrac{1}{15 + \cfrac{1}{1 + \cfrac{1}{292 + \cfrac{1}{\phantom{x}}}}}}$$

come from cutting the continued fraction off at 1, 2, 3, 4 and 5 levels to arrive at:

$3$, $^{22}/_7$, $^{333}/_{106}$, $^{355}/_{113}$, $^{103993}/_{33102}$.

These approximations zoom in on the true value of $\pi$ very quickly, differing by around 0.141, 0.001, 0.0008, 0.0000003 and 0.0000000006. The good approximations to $\pi$ emerge because large numbers appear in $\pi$'s continued fraction. If the numbers that appear in the continued fraction are bounded (that is, never growing larger than some fixed number), this is termed **badly approximable** and, in some sense, is a measure of how irrational the number is.

**No such patterns can be found in the standard continued fraction of $\pi$.**

$$\phi = 1 + \cfrac{1}{1 + \cfrac{1}{1 + \cfrac{1}{1 + \cfrac{1}{1 + \cfrac{1}{1 + \cfrac{1}{1 + \cfrac{1}{1 + \cfrac{1}{1 + \cfrac{1}{\ddots}}}}}}}}}$$

The continued fraction of $\phi$. This pattern makes $\phi$ the most irrational of irrational numbers.

The decimal expansion of $\phi$:

$$\phi = 1.61803\ldots$$

Rational approximations of $\phi$:

$$\frac{1}{1},\ \frac{2}{1},\ \frac{3}{2},\ \frac{5}{3},\ \frac{8}{5},\ \frac{13}{8},\ \frac{21}{13},\ldots$$

A beautiful pattern emerges in the continued fraction for $\phi$ that is not apparent in its decimal expansion. It only ever has 1s in its continued fraction, making it the most irrational of the irrational numbers.

# 4.10 Who invented calculus?

**The invention of calculus was followed by a bitter dispute between its two main protagonists.**

The invention of **calculus** was largely down to two men: the Englishman Isaac Newton (1643–1727) and the German Gottfried Wilhelm Leibniz (1646–1716) – both mathematical geniuses.

Newton published his version of calculus in 1687, in his monumental work *Philosophiæ Naturalis Principia Mathematica*. Three years earlier, in 1684, Leibniz had published his first paper on the subject. Newton claimed, however, that he had already worked out the central ideas of calculus all the way back in 1666, at the tender age of 23. The problem was not who got there first – no one doubted that Newton did – but whether Leibniz had developed his ideas independently.

There is some evidence to suggest that Leibniz saw Newton's work and that he later tried to hide this fact. Newton is not entirely innocent either. The Royal Society investigated the dispute and, in 1713, published a report in Newton's favour – a report written by Newton himself!

Today, historians agree that Leibniz discovered calculus independently from Newton. In one sense, Leibniz even won the dispute, since the notation he invented to express his calculus was so accessible that we still use it today.

**In addition to maths, Newton contributed to a wide range of subjects, including optics and astronomy.**

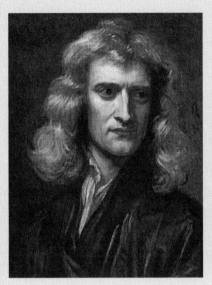

$\dot{x}$   $\ddot{x}$   $\dddot{x}$   $\dot{x}^n$   $\dot{x}^2$   $\dot{x}^3$

$\dfrac{\dot{u}}{\dot{x}}$   $\dfrac{\ddot{u}}{\dot{x}^2}$   $\dfrac{\dddot{u}}{\dot{x}^3}$

$\dfrac{\ddot{u}}{\dot{x}\dot{y}}$   $\dfrac{\dddot{u}}{\dot{x}^2\dot{y}}$   $\left(\dfrac{\dot{y}}{\dot{z}}\right)^{\ddot{}}$

$dx$   $d^2x$   $d^3x$   $d^nx$   $dx^2$   $dx^3$

$\dfrac{du}{dx}$   $\dfrac{d^2u}{dx^2}$   $\dfrac{d^3u}{dx^3}$

$\dfrac{d^2u}{dx.dy}$   $\dfrac{d^3u}{dx^2.dy}$   $d^3\left(\dfrac{dy}{dz}\right)$

Newton (left) with his version of calculus and Leibniz (right) with his.

# CHANCE

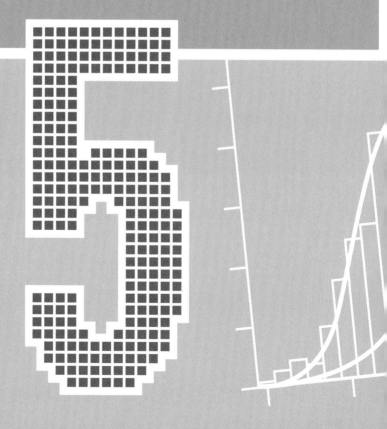

**C** hance is a tricky thing. It's unpredictable by nature, so at first glance it seems impossible to deal with it in any rational way. But it turns out that even chance can be kept in check with a little mathematics.

In this chapter we will find out how to define the probability of something happening, see why your chance of winning the lottery is so incredibly small, and meet the basic laws mathematicians use to calculate with probabilities. We'll explore the concept of randomness and find out how a single number can encode your name, address and entire DNA sequence. We'll discover how infinitely many monkeys typing away on infinitely many typewriters can produce the complete works of Shakespeare.

We'll also see how probability theory is essential in contexts that concern us all. We'll learn how to test the efficacy of a

*Continues overleaf*

drug using randomized controlled trials, and why it is possible to gauge the mood of a nation by questioning a relatively small number of people in an opinion poll. We'll find out how scientists quantify the inevitable uncertainty of their results using significance levels and confidence intervals.

We'll also see how to make mischief with statistics, mixing up different ways of talking about risk to make undesirable outcomes seem smaller than beneficial ones. Finally, we will visit that staple of any crime drama – DNA testing – and learn why matching a DNA sample found at a crime scene doesn't necessarily spell guilt beyond reasonable doubt.

# Contents

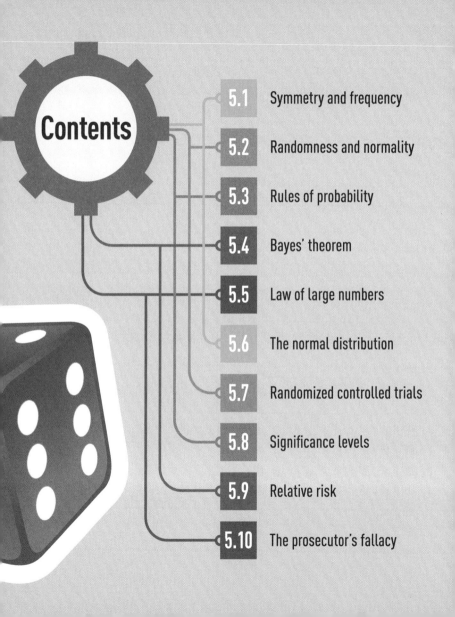

# 5.1 Symmetry and frequency

**The probability of tossing heads on a fair coin is half. But how do we know this?**

One way of defining the probability of tossing heads on a coin is to consider the symmetry of the coin. If it is perfectly symmetrical, then no side is more likely to come up than the other, so the probability of tossing heads or tails should be equal. Moreover, there is no other possible outcome than heads or tails.

Out of a whole of one, therefore, this gives a probability of half to heads and a probability of half to tails. By the same reasoning, the probability of tossing any of the six numbers on a perfectly fair dice is one in six.

But not all processes in the world are symmetrical. This leads to another way of assessing probabilities. Repeat the process (for example, of tossing a coin) a large number of times and work out the proportion (also called the **relative frequency**) of times a certain outcome (say, heads) happens. The idea behind this **frequentist** view of probability is that this proportion approximates the 'true' probability of the outcome. This idea is often used in science. When your doctor tells you that you have a 5% chance of getting a disease, that's because, out of a group of people like you, 5% did get the disease.

The probability of guessing one of the 13,983,816 possible outcomes of a lottery draw is 1/13,983,816 – around 1 in 14 million.

On a perfectly fair dice each of the six sides is as likely to come up as any other. The probability of rolling any of the six numbers is therefore $\frac{1}{6}$ (1 divided by the six equally possible outcomes).

# 5.2 Randomness and normality

**Although we have an intuitive idea of what randomness is, it is surprisingly difficult to define it mathematically.**

We say something is random if it is unpredictable – such as the outcome of flipping a fair coin. However, if you flip a coin ten times in a row you are just as likely to throw ten heads as any other specific combination of ten throws.

This idea led to the first attempt to define randomness mathematically, in 1909. Émile Borel (1871–1956) described a number with an infinite decimal expansion as **normal** if all the digits appeared with the same frequency (1/10th of the time), all pairs of digits with the same frequency (1/100th of the time) and so on. If you decided the digits of a number by tossing a ten-sided dice, the resulting number would be normal. Normality is used as a test for randomness – if a number isn't normal, its sequence of digits isn't considered random.

Borel showed that most numbers are normal, but couldn't give an actual example of one. An undergraduate, D. G. Champernowne finally produced the first example in 1933:

0.12345678910111213141516171819202122 23 . . .

**For a number to be random it needs to be normal, but not all normal numbers are random.**

This number lists the digits of every whole number after the decimal place. As well as not favouring any combination of digits, normality also means every combination should appear. So within this, and every other normal number, your age, phone number and DNA sequence are encoded in numerical form.

# Infinite monkey theorem

We have Borel to thank for the idea that led to the infinite monkey theorem – that an infinite army of monkeys typing on typewriters would eventually produce the complete works of Shakespeare.

# 5.3 Rules of probability

**Probability may be a somewhat slippery concept, but in maths there are strict rules for working with it.**

In maths, the probability of an event happening is always taken to be a number between 0 (no chance of it happening) and 1 (it will happen for sure).

Some basic rules apply. If two independent events A and B have probabilities P(A) and P(B), respectively, then the probability of either A or B occurring is:

P(A or B) = P(A) + P(B).

The probability of both A and B occurring is:

P(A and B) = P(A) × P(B).

So, if A is the event of throwing a 2 on a fair dice and B the throwing of a 3, then the probability of throwing either a 2 or a 3 is: P(2 or 3) = P(2) + P(3) = 1/6 + 1/6 = 1/3.

The probability of throwing a 2 and a 3 on two rolls of the dice is: P(2 and 3) = 1/6 x 1/6 = 1/36.

Both of these rules make sense: 2 and 3 together form one-third of the collection of all possible outcomes of one roll (1 to 6), so the probability of throwing one of them should indeed be 1/3 (see Topic 5.1). With two rolls there are 6 × 6 = 36 possible outcomes. Rolling a 2 and a 3 is one of those outcomes, so P(2 and 3) should indeed be 1/36.

**The probabilities of mutually exclusive outcomes of a process always sum to 1.**

## Probabilities of two dice rolled together

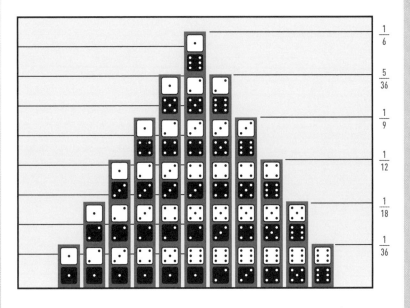

$$2 \times \frac{1}{36} + 2 \times \frac{1}{18} + 2 \times \frac{1}{12} + 2 \times \frac{1}{9} + 2 \times \frac{5}{36} + \frac{1}{6} = 1$$

This pyramid of dice demonstrates probability when throwing two dice. The chances of throwing two 1s or two 6s, for example, are 1 in 36, the chances of throwing a 5 and a 6 are 1 in 18, and so on. The sum of the probabilities of the different outcomes (shown above) is equal to 1.

# 5.4 Bayes' theorem

It would be foolish to ignore evidence. Thankfully Bayes' theorem allows us to update our beliefs when new evidence becomes available.

A particular type of disease affects 1% of the population. There is a test for the disease, but it's not perfect: it gives a positive result for 90% of people who have the disease, but also for 5% of the people who are disease-free. You have just received a positive test result – what is the probability that you have the disease? Many would say 90%, but actually your chances are closer to 15%.

A **conditional probability** is the probability that one event, A, happens, given another event, B, has already happened: written $P(A \mid B)$. **Bayes' theorem**:

$$P(A \mid B) = P(A) \times P(B \mid A)/P(B)$$

allows you to calculate with conditional probabilities, updating your original belief about the probability of A, given the new evidence of B, that you've had a positive test result: $P(\text{disease} \mid \text{positive})$.

Our belief that we had the disease before the test was $P(\text{disease})=0.01$. You can calculate $P(\text{positive})$ to be 0.0585 (this combines the proportion of sick people who correctly test positive, and the proportion of healthy people who incorrectly test positive). We also know that $P(\text{positive} \mid \text{disease}) = 0.9$. Using these figures, Bayes' theorem says:

$$P(\text{disease} \mid \text{positive}) = P(\text{disease}) \times P(\text{positive} \mid \text{disease})/P(\text{positive}) = 0.01 \times 0.9/0.0585 = 0.154.$$

Bayes' theorem is named after Thomas Bayes (1701–61), who was a Presbyterian minister as well as a statistician.

## Demonstrating Bayes' theorem

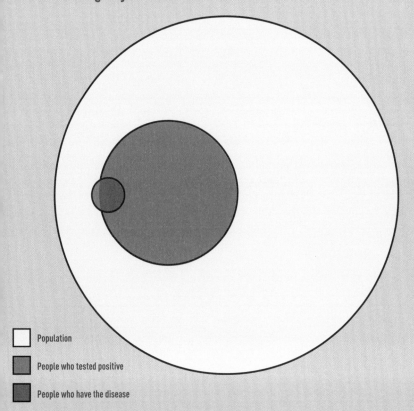

Population

People who tested positive

People who have the disease

The intersection of the green and orange circles represents the proportion of people who have the disease and a positive test result, out of all the people who've had a positive test result.

# 5.5 Law of large numbers

**The law of large numbers is something we all know about intuitively, but often misunderstand.**

The law of large numbers says that repeating a process a large number of times will result in outcomes that reflect the underlying probabilities. For example, when flipping a fair coin many times, the proportion of heads should be roughly half, reflecting the fact that the probability of heads in a single flip is half.

This idea is often misunderstood. If you have just flipped the coin 99 times and got heads at each flip, doesn't the law of large numbers imply there's a very high chance to get tails at the 100th flip? After all, the law of large numbers implies the proportion of tails should be as close as possible to half.

The answer is no. The chance of flipping tails at the next flip is still half. The law of large numbers says that, as the number of your coin flips grows towards infinity, the proportion of heads will converge to 0.5 (see Topic 4.2). But this allows for the possibility that the proportion of heads only gets anywhere near 0.5 after, say, the millionth, billionth or trillionth flip. It is under no obligation to get near to 0.5 on the 100th flip.

**Don't rely on the law of large numbers to decide the next call on a coin toss.**

It's a long way to infinity, and a very short way to losing money if you're not careful!

## Heads or tails?

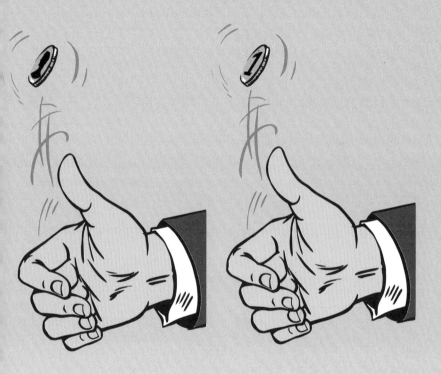

Every time you flip a coin you have exactly the same chances of
throwing heads as you do throwing tails.

# 5.6 The normal distribution

A central idea in statistics is that you can say something about a whole population based on a smaller sample. But how do we know that this actually works?

Suppose you have computed the average height of a random sample of 30 people, and found that it's 75 cm (2 ft 6 in)? You'd know that something funny has happened: somehow your sample included lots of very short people. But how do you deal with the fact that such an unusual result could occur in any study that doesn't involve the whole population?

The answer comes from a near-miraculous fact. Suppose you took random samples from lots of groups of 30 people, computed the average for each group and noted how many times each average occurs on a frequency plot. Then, no matter what you are sampling, be it height, approval rankings or income, the frequency plot will always look approximately like a bell-shaped curve. The more people in each of your samples, the closer your plot approximates a bell curve. The top of the bell, which corresponds to the average you observe most often, is the true average in the whole population that you are seeking.

Knowing that averages from lots of samples are distributed in this way enables statisticians to work out how far the average of a single sample – for example, a single opinion poll or height sample – is likely to be from the true average they are seeking, thereby assessing their confidence in the estimate.

**Bell curves are examples of the normal distribution and the miraculous fact is the central limit theorem.**

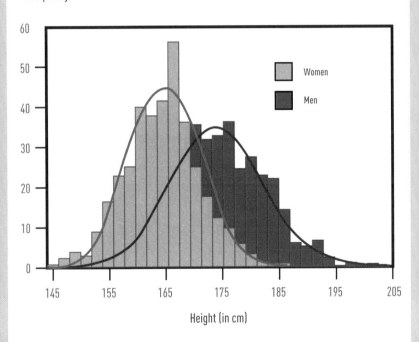

## Average height

Frequency

The frequency plot of average heights approximates a normal distribution. Many quantities we observe in the real world are averages of processes we can't see, so the normal distribution crops up often.

# 5.7 Randomized controlled trials

**How do you decide if a medical treatment is effective?**

Historically, doctors decided which medical treatments worked by trial and error, based on personal experience. But this is unreliable, as it is not possible to tell if a patient gets better (or worse) owing to your treatment, or some other factor that you don't know about. So, in the 19th century, scientists began testing treatments using two groups of people, the **study group** – those who received the new treatment – and the **control group** – those who were given either an inert treatment (a **placebo**) or an existing treatment that was being compared to the new one.

People tend to see what they are expecting to see, however. For this reason, in 1917, the concept of **blinding** was introduced, where the people involved in medical trials didn't know whether a patient was in the control or study group. Now the knowledge of which group a patient is in couldn't interfere (intentionally or unintentionally) with the results of the tests.

**Randomized controlled trials are now universally used to evaluate new medical treatments.**

There was still a way to influence the results, however, by preferentially giving your new treatment to patients who were less unwell, and so were likely to have a better outcome. In the 1940s, therefore, the trial of whooping cough vaccines became the first **randomized controlled trial** in which patients were allocated *randomly* to each of the study and control groups.

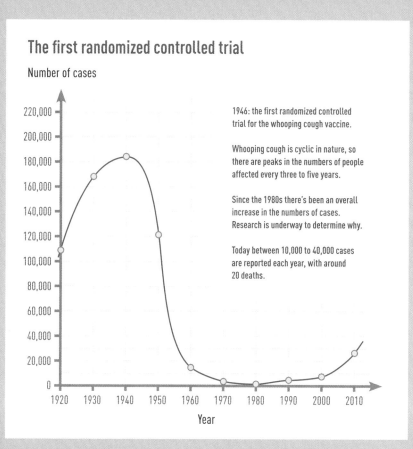

## The first randomized controlled trial

Number of cases

1946: the first randomized controlled trial for the whooping cough vaccine.

Whooping cough is cyclic in nature, so there are peaks in the numbers of people affected every three to five years.

Since the 1980s there's been an overall increase in the numbers of cases. Research is underway to determine why.

Today between 10,000 to 40,000 cases are reported each year, with around 20 deaths.

Year

This graph represents cases of whooping cough since 1920. Before the vaccine became available in the 1940s, around 200,000 children became sick, with about 9,000 of those dying, in the US each year.

# 5.8  Significance levels

**When scientists report a result, they often talk about *p*-values and significance levels. What do these things mean?**

Suppose in a randomized controlled trial (see Topic 5.7) a new drug reduces blood pressure by an average of 20mmHg (the standard unit of blood pressure). How do you know that the reduction in blood pressure really occurred because of the drug, and not for all sorts of other reasons that we don't know about?

While you can't work out the probability of the drug being effective, you *can* work out the probability of seeing the difference as big as the one you observed (a reduction of 20mmHg), assuming the drug is not effective. This is known as the ***p*-value**. If the *p*-value is small, then that's clearly a good reason to believe that the drug is effective.

So how small is small? Typically, in medical studies, a *p*-value of less than 5% is taken to indicate a significant result. This threshold of 5% is known as the **significance level**, usually denoted by α (the Greek letter alpha). We say a trial is significant at level α if there was a probability less than α of seeing the result (a reduction of 20mmHg) if the drug were not effective, and the result was due to chance alone.

The *confidence interval* is a related concept. For example, the interval from 15 to 25 is called a 95% confidence interval if you are 95% confident that the true result (the true reduction in blood pressure) lies between 15 and 25.

**A 95% confidence interval corresponds to a significance level of 5%.**

# Testing drugs

This graph illustrates a distribution measuring the probability that a test result is equal to *x* if the drug is not effective. The probability that the result corresponds to the red regions is 5%.

Test result

Probability density

When testing drugs such as vaccines we need to exclude the possibility that the result of a study was due to chance.

# 5.9  Relative risk

**Put down those bacon strips – bacon increases your risk of bowel cancer by 20%! But is this favourite breakfast really as guilty as it sounds?**

A report from the World Cancer Research Fund said, among many other things, that eating 50 g (2 oz) of processed meat every day (equivalent to a bacon sandwich), increases your chances of getting bowel cancer by 20%.

If this sounds alarming, it is because it is stated in terms of **relative risk** – that is, how much your risk increases relative to the **absolute risk** (how many people in the general population can be expected to get the illness).

Around 5% of the population get bowel cancer, and 20% more than 5 is 6. So the relative risk of 20% from daily bacon sandwiches translates as increasing your absolute risk to 6%. An increase in absolute risk of 1% is important to consider, but doesn't sound quite as alarming as the increase of 20% in terms of relative risk.

Relative risks sound larger than absolute risks. It is, therefore, possible to spin the results of a test by reporting relative or absolute risks selectively. For example, if you want positive results for a new drug, you might report relative risks for its effectiveness, but absolute risks for any unpleasant side effects. This way, the effectiveness of the drug appears to outweigh its side effects. This is called **mismatched framing**.

**The spurious practice of mismatched framing was found in one-third of studies published in important medical journals.**

## Increased risk

We are told that eating 50 g (2 oz) of bacon daily increases your chances of getting bowel cancer by 20%. This sounds alarming, but how worried should you really be?

# 5.10 The prosecutor's fallacy

A woman's DNA matches that of a sample found at a crime scene. The chances of a DNA match are just one in two million, so the woman must be guilty, right?

Wrong. But it's a common mistake to make, known as the **prosecutor's fallacy**. It mistakes the one in two million for the probability of the woman's innocence. In order to assess the woman's guilt properly, we need to take the fact that she matched the sample as a given, and see how much more likely this makes her to be guilty than she was before the DNA evidence came to light.

A version of Bayes' theorem (see Topic 5.4), stated in terms of gambler's odds, is useful here. The matching probability above implies that the woman's DNA is two million times more likely to match the sample if she is guilty, than if she is innocent. Bayes' theorem now says that:

Odds of guilt after DNA evidence = 2,000,000 × Odds of guilt before DNA evidence.

If our woman comes from a city of 500,000 people, and we think each of them is equally likely to have committed the crime, then her odds of guilt before the DNA evidence are about 1 in 500,000. Therefore:

Odds of guilt after DNA evidence = 2,000,000 × 1/500,000 = 4.

Translating into probabilities, this gives an 80% chance of guilt. Definitely not beyond reasonable doubt!

There are recorded examples of the prosecutor's fallacy leading to wrongful convictions.

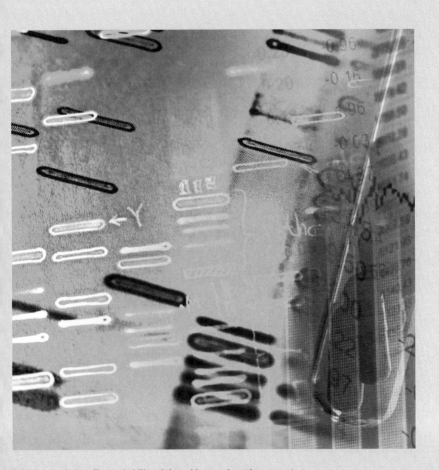

DNA fingerprinting. The probability of the evidence, given the suspect is innocent, is not the same as the probability of the suspect being innocent, given the evidence.

# CURVES

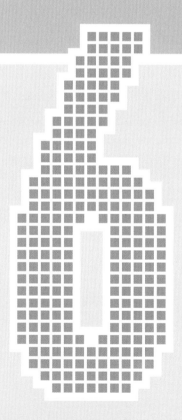

I n 1915 Albert Einstein changed the way we think about the universe. He realized that the force of gravity is the curvature of spacetime. Curves also provide some of the most beautiful – and powerful – illustrations of mathematics. Their shapes can reveal the nature of the equations that define them, and they provide the machinery with which to explore some of the most complicated mathematical objects.

Einstein wasn't the first person to use curvature to shed light on the workings of the universe. In the 17th century, curves called ellipses provided Johannes Kepler with the shape of planetary orbits. The ellipse itself was first studied by the ancient Greeks as one of the conic sections. Conic sections are the curves created by taking a flat slice through two cones balanced on their points, to give us circles, ellipses, parabolas and hyperbolas. Another favourite curve from the 17th century

*Continues overleaf*

was the catenary, which Robert Hook realized was the shape of a perfectly efficient self-supporting arch.

Curves enable us to study curvature, whether of intricate curves or complex surfaces. The first step to understanding curvature is that any curve can be approximated by a flat tangent line. Curves can also be approximated by the beautifully named osculating circles or kissing circles. Both of these concepts are then used to move beyond curves in order to calculate the curvature of surfaces.

Minimal surfaces – the delicate shapes created by soap films spanning a wire frame – are characterized by their curvature. Architects call on this field of mathematics for its efficiency with matter as well as the beautiful shapes.

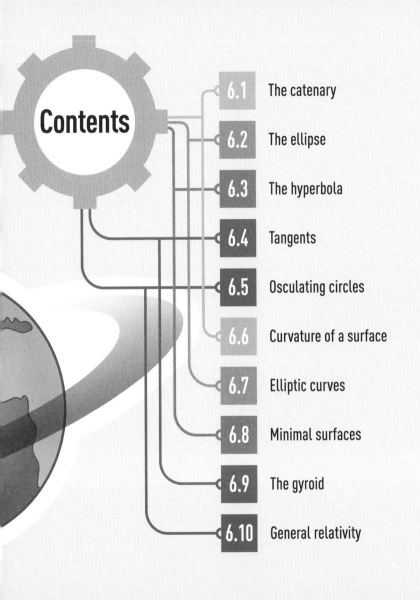

# Contents

# 6.1 The catenary

**What do Wembley stadium and St. Paul's cathedral in London, have in common?**

When you suspend a chain from two hooks and allow it to hang naturally under its own weight, the curve it describes is called a **catenary**. Any hanging chain will find this equilibrium shape, in which the forces of tension (coming from the hooks holding the chain up) and the force of gravity pulling downwards are in perfect balance.

Something beautiful happens when you turn a catenary curve upside down. The inverted shape now describes an arch – in fact, the most stable shape an arch can take. In a hanging chain, the forces of tension act along the line of the curve. In the inverted catenary the forces of tension become forces of compression, and since these forces are also directed along the line of the arch, the shape doesn't bend or buckle. For this reason, anyone wanting to build an arch should make sure it contains the shape of an inverted catenary. That way it will stand freely under its own weight. Not only that, but building this arch will use the minimum in terms of materials.

**The shape of the Gateway Arch in St. Louis, USA, the world's tallest arch, is based on an inverted catenary.**

The British architect Robert Hooke (1635–1703) was the first person to study the catenary mathematically. In 1675 he published an anagram (in Latin) of: 'As hangs the flexible line, so but inverted will stand the rigid arch'.

# The dome of St. Paul's cathedral, London

St. Paul's has three domes: an outer dome and an inner dome for visual effect, and a hidden middle dome for strength. Sir Christopher Wren based the shape of the middle dome on an inverted catenary.

# 6.2 The ellipse

**A circle is a special case of another type of curve – the ellipse.**

You can draw a circle using a pencil and a loop of string: the pencil is at one end of the loop, with the other end held at the centre. Pulling the loop tight and tracing around the centre describes the circle.

You can use the same loop of string to draw an **ellipse**. This time the loop is stretched between three points – the pencil and two other points that are called the **foci** (plural of focus) of the ellipse. This time, when the loop is pulled tight it forms a triangle, with one side always remaining the same length (the space between the fixed foci). The two remaining sides (the distances from a point on the ellipse to each of the two foci) always add up to the same length. You can try this with a pencil and a loop of string, using your thumb and index finger as the fixed foci.

A circle is just a special type of ellipse where the two foci are actually the same point. Otherwise, as the foci move apart, the ellipse elongates. This stretching is measured by the **eccentricity** of the ellipse: the ratio of the distance between the foci divided by the width of the ellipse.

**If you slice a traffic cone with a slightly angled plane, the cross-section is an ellipse (see page 139).**

## Kepler's laws

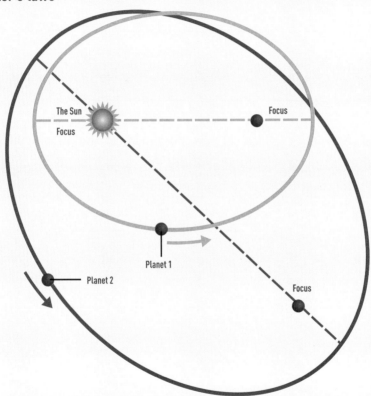

In the 17th century, Johannes Kepler discovered three laws that described the motions of the planets. The first law was that the orbit of each planet is an ellipse, with the Sun at one focus. The dotted lines illustrate how the axis of each ellipsis runs through both of its foci.

# 6.3 The hyperbola

**This curve belongs to an important family of curves that have been studied for millennia.**

If you shine a torch straight down at the floor, you'll create a circular pool of light, as the floor intercepts the cone of light. Tilt the torch, and your circle will elongate into an ellipse. Tilt it further, and the ellipse will stretch away, until the upper side of the torch's light cone is parallel to the floor. Your light pool will now be the shape of a **parabola** (see Topic 3.3). Tilting the flashlight further still produces a shape known as a **hyperbola**.

These shapes – circles, ellipses, parabolas and hyperbolas – have been studied mathematically since 300 BC, and are called the **conic sections**: slices through two cones that are placed one above the other, tip to tip (see opposite). A horizontal slice gives a circle, an angled slice gives an ellipse, a slice parallel to the side of the cone gives a parabola, and a slice that cuts through the two cones gives you a hyperbola.

A hyperbola has two arms, each an exact mirror image of the other. These arms are contained by two, crossing, straight lines called **asymptotes**. As the arms run away from the middle of the hyperbola, they get closer and closer to these lines, but never touch them.

**You've probably already met a hyperbola – the graph of the function $y = 1/x$.**

## The conic sections

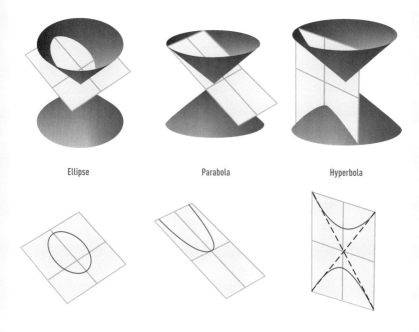

Ellipse        Parabola        Hyperbola

The images above show the three conic sections: the ellipse (which includes the circle as a special case), the parabola and the hyperbola. The hyperbola is contained within its two asymptotes, shown as dotted black lines.

# 6.4 Tangents

**Zoom right in on a curve that's been drawn on a flat piece of paper and, at any one point, it will look like a straight line.**

A straight line that just touches a curve at a point is called a **tangent** to the curve at that point. If a curve is smooth, it must have a tangent at every point. Something with a sharp corner, such as the absolute value function, has tangents at almost all points, except at the corner. For a tangent to exist, its direction must smoothly change as you go from one side of the point to the other. Whereas, at the corner of the absolute value function, the tangent shifts direction abruptly. The two are demonstrated opposite.

You can also have a tangent line, and a tangent plane, to a surface. The same rules apply: a smooth surface has tangent lines and tangent planes defined at every point – no kinks or creases allowed.

The curvature of a curve or a surface is measured by how much it differs from being flat. For a curve, this means how well it is approximated by the tangent at any point. For a surface, this means how well it is approximated by its tangent plane at any point. If a curve is very similar to its tangent at a point – say, at large positive or negative values for a parabola, the curvature is very close to zero (the curvature of a straight line). But if the curve is very different from its tangent, say at the turning point of the parabola, the curvature will have a larger positive, or negative value.

**A line perpendicular to the tangent line (for a curve) or tangent plane (for a surface) is called a normal line.**

## Tangents and curvature

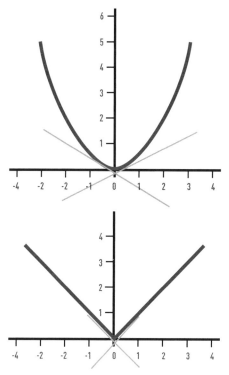

Curvature is the rate of change of direction of the tangents as you move at a steady speed along the curve. For the absolute value function (lower image) the tangent is unchanging before and after the corner, but makes an instantaneous change at the corner – which isn't allowed.

# 6.5 Osculating circles

**Curvature determines how different a curve is from a flat line. But how is this measured?**

Tangent lines are straight lines that approximate a curve very close to a particular point. Another way to approximate a curve at a point is with a circle. The **osculating circle** at a point on the curve is the circle that most closely matches the curve at that point. The size of the curvature at that point is the value of $1/R$, where $R$ is the radius of the osculating circle.

If the curve is relatively flat – say, at very large negative or positive values of $x$ for the curve $x^2$, then the osculating circle will have a very large radius and the curvature $1/R$ will be close to zero. If the curve is much more pronounced – say, at the turning point of the parabola for $x^2$, then the radius of the osculating circle will be much smaller, and the curvature will be much further from zero.

The curvature can be defined as negative or positive, depending on which side of the curve the osculating circle lies. We can assess the overall curvature of a curve by looking at the rate of change of the curvature when moving at a steady speed along the curve. For example, a circle has constant curvature – that is, the osculating circle is just itself at every point, which remains unchanged as you move around the circle.

**Osculating circle means 'kissing circle', as the circle just touches, or 'kisses', the curve at that point.**

# Osculating circle for a sine wave

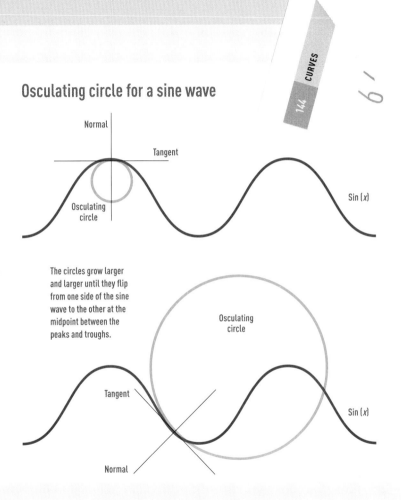

Normal

Tangent

Osculating circle

Sin (x)

The circles grow larger and larger until they flip from one side of the sine wave to the other at the midpoint between the peaks and troughs.

Osculating circle

Tangent

Sin (x)

Normal

The osculating circle for a sine wave shrinks and grows as you move along the curve. The smallest circles are at the points of greatest curvature – the peaks and troughs.

# Curvature of a surface

## How do we measure the curvature of a surface?

Osculating circles give a numerical value for curvature at a point on a one-dimensional curve drawn on a flat plane (see Topic 6.5). We can use the same method to calculate the curvature of a two-dimensional surface.

The curvature of a surface is how much the surface differs from its tangent plane at this point. The **normal** to the tangent plane is the direction perpendicular to it. You can define a plane that contains this normal direction, making a **normal plane** perpendicular to the tangent plane. The surface slices through the normal plane, creating a one-dimensional curve on this flat plane, the curvature of which can be measured by the osculating circle.

There are infinitely many normal planes at any point on the surface, each giving a different value of the curvature for the curves sliced out of these normal planes by the surface. We need only to consider the maximum and minimum values of these. The product of these maximum and minimum curvatures calculated from the normal planes is called the **Gaussian curvature**. A surface is flat at a point if this product is zero. A surface has positive curvature if the Gaussian curvature at that point is positive – at that point the surface would be similar to a bowl or a hill. Instead, if the surface is saddle-shaped at a point, the Gaussian curvature is negative.

**There are more complicated ways to define curvature for the higher-dimensional analogues of surfaces.**

## Gaussian curvature

Two normal planes showing the maximum and minimum curvatures of the surface at a given point.

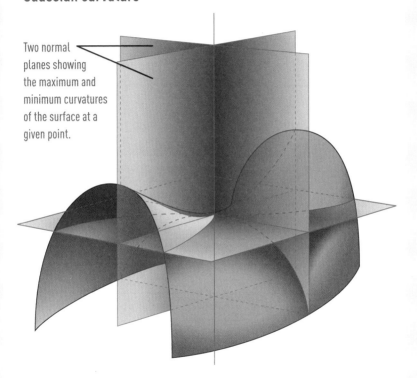

The Gaussian curvature of this saddle-shaped surface is negative. It is the product of the maximum curvature and minimum curvature, which have opposite signs, as demonstrated by the image: the curves in the normal planes lie on different sides of the tangent plane.

# 6.7 Elliptic curves

**Can you find two whole numbers $x$ and $y$ so that $y^2 = x^3 - 2x + 1$? The answer is yes, for example, $x = 0$ and $y = 1$ will do. But are there other solutions?**

If you plot all the points $(x, y)$ that satisfy the equation:

$$y^2 = x^3 - 2x + 1$$

in a coordinate system (see Topic 3.2), the result is the beautiful curve framed in the diagram opposite. This is an example of an **elliptic curve**.

Our elliptic curve encapsulates, at a single glance, the pairs of real numbers $x$ and $y$ that are solutions to our equation. But what if we wanted to see all pairs of **complex numbers** (see Topic 1.9) that are solutions, too? A complex number comprises two pieces of information, so a pair of complex numbers comprises four pieces of information. To plot the pair, therefore, we need four dimensions, which we cannot visualize.

There is, however, a useful fact that comes to our rescue. The shape encoding the complex solutions to our equation does *live* in four-dimensional space, but it only actually uses up two of these dimensions. In fact, it can be represented by a surface we can easily imagine: a doughnut. The same is true for every equation that defines an elliptic curve. The pairs of complex numbers that are solutions to the equation can always be represented by a doughnut.

**Elliptic curves played an important role in the proof of Fermat's last theorem.**

## Elliptic curves

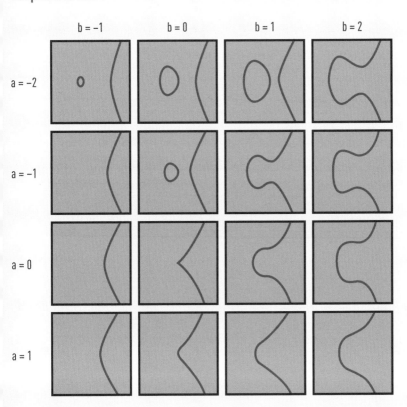

An elliptic curve is described by an equation of the form
$y^2 = x^3 + ax + b$. The above diagram shows the elliptic curves
we get when $a$ ranges from −2 to 1 and $b$ ranges from −1 to 2.

# 6.8 Minimal surfaces

Good news: we can all solve complex maths problems while playing with soap bubbles.

A circle is the most efficient shape in that it encloses the largest area for a given length of perimeter (see Topic 2.3). Equally, you could say that a circle is the shape with the shortest perimeter enclosing a given area. And the same goes for a sphere, it minimizes the surface area required to enclose a given volume of space.

You can ask similar questions, such as: how can you divide up a plane into equal-sized pieces, so that the dividing perimeters have the shortest length? Mathematicians have known since at least AD 360, that the answer to this is a hexagonal honeycomb. However, they only managed to prove this Honeycomb Conjecture in 1999.

Nature provides answers without the mathematical proof, because nature – whether relating to bees or bubbles – arrives at a solution that requires the minimal amount of energy. Mathematically, a **minimal surface** is one that has the smallest possible area while spanning some boundary. You could find a minimal surface by solving the corresponding equations. Or, more easily, you could twist a piece of wire into the shape of the boundary you want, and dip it into soapy liquid. The soap film spanning your wire when it emerges will be the same as the mathematical solution.

**Mathematically, minimal surfaces are those with zero Gaussian curvature (see Topic 6.6) – meaning that spheres aren't technically minimal surfaces.**

Minimal surfaces appeal to architects. Frei Otto famously used
the minimal surfaces produced by soap films to aid his design
for Munich's Olympiapark (1972).

# 6.9 The gyroid

**A labyrinthine surface that creates beautiful colours and stubborn sauce.**

Interest in minimal surfaces (see Topic 6.8) was revived in the 1970s thanks to a discovery made by Alan Schoen, a scientist working at NASA on superlight structures. Schoen discovered the **gyroid**, a minimal surface that divides space up into two twisting labyrinths. A particularly special feature of the gyroid is that it has a regular repeating shape – it is built from a fundamental piece that repeats in each of three directions.

The gyroid may be hard for us to picture, but this highly complex shape has been found repeatedly in nature. The beautiful iridescence of a butterfly's wing is caused by periodic variations in the light that is scattered by the small scale structures in the surface – and in several species this regular structure has been observed to have the shape of a gyroid. Gyroids have also been observed within cell membranes, and within certain kinds of plastics and rubber.

**The ability to assemble gyroids out of a fundamental piece makes it an area of interest in nanotechnology.**

Gyroids may also explain why ketchup can be so hard to get out of a glass bottle. Imperfections in the regular gyroid shape have a large effect on the flow, resulting in the stubbornness of this condiment.

## Building blocks of the gyroid

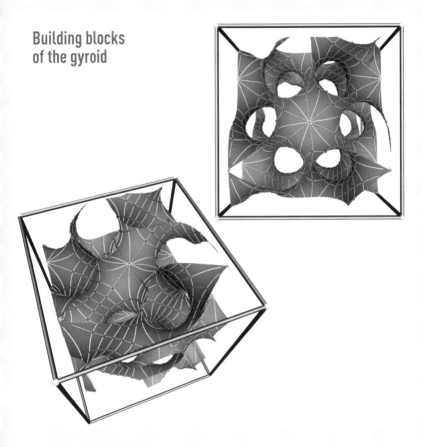

The gyroid can be assembled from a fundamental piece, that infinitely repeats in three directions. The surface this produces separates space into two intertwining labyrinthine passages.

# 6.10 General relativity

In 1915, Albert Einstein's general theory of relativity changed the way we view the universe: gravity isn't a force, it's the curvature of spacetime.

Most of us learn about gravity as it was described by Isaac Newton (1642–1727) in the 17th century: two objects exert a gravitational force on one another that is proportional to the product of their masses divided by the square of the distance between them.

This is a good description that gives answers that match our experience of gravity in everyday life. This description also satisfied science for more than two centuries. But in one thought experiment, Albert Einstein (1879–1955) realized that Newton's theory couldn't be correct. Imagine the Sun exploded. Owing to the time it takes the Sun's light to reach the Earth, we wouldn't see this disaster for eight minutes. However, according to Newton's theory, we should instantly feel the loss of the Sun's gravity.

This is a problem because in 1905, in his special theory of relativity, Einstein had shown that nothing travels faster than light, including information. The instantaneous loss of the Sun's gravity would have contravened this.

Instead, Einstein changed the way we viewed both space and gravity. He told us that we shouldn't make a distinction between space and time. Instead, we should think of them as two parts of the same concept: **spacetime**. And gravity is the result of spacetime being curved by matter.

'Space and time are modes by which we think, and not conditions in which we live.'
Albert Einstein

## Warping spacetime

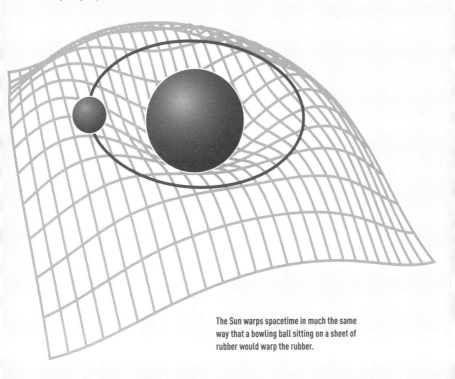

The Sun warps spacetime in much the same
way that a bowling ball sitting on a sheet of
rubber would warp the rubber.

The massive Sun warps spacetime. The curve it creates means that an
object moving towards it will roll down the curve, moving around the
Sun like a marble circling in a bowl.

# PATTERNS
# AND SYMMETRY

**T**he famous number theorist G. H. Hardy (1877–1947) once said: 'A mathematician, like a painter or a poet, is a maker of patterns. If his patterns are more permanent than theirs, it is because they are made with ideas.'

This quote by Hardy captures what many mathematicians feel about their art: that mathematics is the language of patterns and forms. Some of these patterns, like the beautiful symmetry of a snowflake, are visible, while others are hidden from view. They lie concealed within the mathematical structures that describe the world around us.

In this chapter we explore both the visible and the hidden. We start with the concept of symmetry, which every child can grasp and which some people think is a prerequisite for beauty. We find out why there are only so many ways you

*Continues overleaf*

can tile your bathroom, and why there's
a fundamental limit to the number of
wallpaper patterns you can choose for
your living room. We also learn about
one famous mathematician's quarrel
with Kleenex, over a clever way of
patterning toilet paper based on
tilings that never repeat.

Next, we explore the hidden symmetries
that turn up in equations, mathematical
problems and other abstract structures.
We find out how those symmetries
can help you solve a problem, and
how symmetries are related to
fundamental laws of physics.
We also learn about the abstract
study of symmetry, which involves
two of mathematics' most tragic
heroes (see page 70), and see how
it has led to the biggest mathematical
proof in history.

# Contents

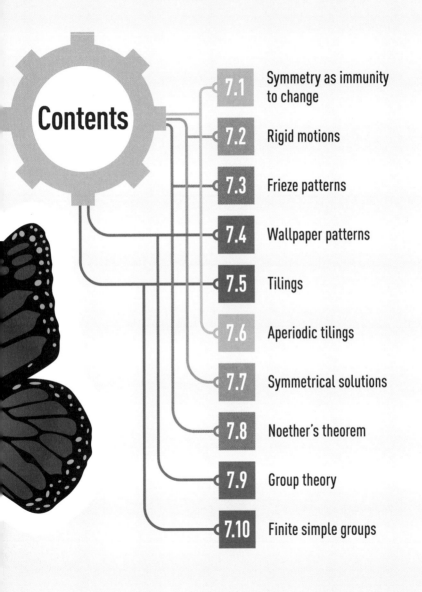

# 7.1 Symmetry as immunity to change

**We have an intuitive understanding of symmetry, while maths has a very precise definition.**

You have a square pinned at its centre to a sheet of paper. Close your eyes and rotate the square by 45 degrees – one-eighth of a turn. Open your eyes, and you'll be able to spot the difference. But rotate the square by 90 degrees – a quarter turn – and the shape will look unchanged.

A **symmetry** is an operation you can perform on an object that essentially leaves it unchanged. The square in our example is immune to rotation by 90 degrees, but is changed by a rotation of 45 degrees. This is because it has **four-fold rotational symmetry** but doesn't have **eight-fold rotational symmetry**.

Symmetry can take many forms and is found in many different objects. Physical objects can have many different types of symmetry – such as rotational and reflective. But mathematical objects, even equations, can also exhibit symmetries, such as immunity to changes in variables.

Humans are said to have an innate appreciation of symmetry. We find it easy to spot and there are even experiments investigating whether we prefer it (finding symmetrical faces more beautiful to asymmetric ones). But a lack of symmetry also has an important place in aesthetics – our eye is automatically drawn to where symmetry breaks down.

**Stone balls with symmetrical carvings from the Neolithic period in Scotland are an early example of our fascination with symmetry.**

Symmetry is common in nature. Consider the reflective symmetry
that is almost perfectly shown in a tiger's face. But asymmetry is also
important, such as in the handedness of proteins within our bodies.

# 7.2 Rigid motions

**What kinds of things can you do to a piece of paper without distorting a picture drawn on it?**

Clearly you can't fold the paper, crumple it up or stretch it. What you can do, however, is to slide the sheet of paper along by some distance in some direction. Such a transformation is called a **translation**. You could also put your thumb down on the paper, keeping one point of it fixed, and then turn the sheet around that fixed point through some angle. That's a **rotation**. Another option is to draw a line on the paper and reflect the picture in that line, turning it into a mirror image of itself.

Are there other, more complicated, things you could do? The answer is: no. A transformation that preserves distances between points on the plane is called a **rigid motion**. There are only four different kinds of rigid motions: translations, rotations, reflections and **glide reflections**. The last involve a reflection followed by a translation in a direction parallel to the reflection axis.

Rigid motions give us a way of expressing what we usually mean when we say a shape is **symmetric**: it remains the same when we apply a reflection, rotation, translation or glide reflection to it.

**A butterfly has mirror symmetry while a snowflake has six-fold rotational symmetry.**

# Rigid motions

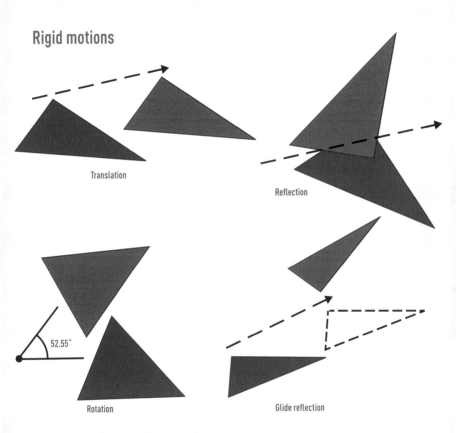

Translation

Reflection

52.55°

Rotation

Glide reflection

Rigid motions give us a way of expressing what we usually mean when we say a shape is symmetric: it remains the same when we apply a reflection, rotation, translation or glide reflection to it.

# 7.3 Frieze patterns

You can characterize patterns by the symmetries they contain. Although two patterns could be made of very different elements, they can exhibit the same types of symmetry.

You can see frieze patterns in ribbons, decorative strips on buildings, even in your own footsteps. A frieze pattern is one that has **translational symmetry** (see Topic 7.2) – that is, you can slide the pattern along a fixed distance, and it will appear unchanged.

Mathematicians characterize frieze patterns by the symmetries they contain. The simplest frieze pattern, known as the 'hop', has only translational symmetry. Other patterns contain many symmetries, such as the 'spinning jump', which has translational symmetry, horizontal and vertical symmetry (remaining unchanged when reflected in a horizontal or vertical line), and rotational symmetry. The pattern that matches our walking footsteps has glide symmetry, where the pattern remains unchanged after you have slid it along and then reflected it in a horizontal line.

There are seven ways to combine these symmetries, resulting in seven frieze patterns. These had all been discovered by artists centuries ago and you can see many in the decoration of ancient buildings. But mathematicians weren't able to prove that there were exactly seven frieze patterns – and no more – until the 19th century.

**The oldest examples of all seven frieze patterns come from the Paleolithic era: 25000–10000 BC.**

## The seven patterns

Hop
(translational symmetry)

Step
(translational and glide symmetry)

Jump
(translational and horizontal symmetry)

Sidle
(translational and vertical symmetry)

Spinning hop
(translational and rotational symmetry)

Spinning jump
(translational, rotational, horizontal
and vertical symmetry)

Spinning sidle
(translational, rotational, glide and
vertical symmetry)

There are seven frieze patterns. They all have translational symmetry,
and some have rotational symmetry, horizontal symmetry, vertical
symmetry and glide symmetry.

# 7.4 Wallpaper patterns

**Given the near infinite variety of wallpaper designs, it is perhaps surprising that mathematicians recognize only 17 different types.**

**At least 14 of the 17 wallpaper patterns have been found at the Alhambra palace in Spain.**

To mathematicians, a wallpaper pattern is one in which a basic block – say, a rose in floral wallpaper – repeats again and again in two different directions. Since you can shift it along in those two directions without changing it, such a pattern has two translational symmetries. But it may also contain reflectional and rotational symmetries, and be symmetric under glide reflections.

While old-fashioned wallpaper may appear very different to the ultramodern pattern you have chosen for your own home, the two may be the same in terms of the symmetries they possess. Investigating all the possibilities, mathematicians have identified a total of 17 **wallpaper groups**, each recording a particular configuration of symmetries. Interestingly, there are some restrictions. For example, rotations in a wallpaper pattern can only have angles of 60, 90, 120 and 180 degrees – you just can't fit together the basic blocks to make up a pattern that gives you any other type of rotational symmetry. It's a phenomenon we'll encounter with tilings (see Topic 7.5).

The Russian mathematician Evgraf Fedorov (1853–1919) proved that there were exactly 17 groups in 1891. Fedorov was also a chemist and this was his motivation for the proof. Chemists use symmetries to understand the behaviour of chemical compounds.

This repeating star pattern can be found in wall panelling at the
Alhambra Palace in Granada, Spain.

# 7.5 Tilings

**If you have ever wondered why most bathroom tiles are square, here's why.**

There are not many regular polygons (see Topic 2.2) that can tile the plane. The only options are equilateral triangles, squares and regular hexagons, which fit together to give the familiar honeycomb pattern.

Try fitting together copies of a regular pentagon and you will soon falter. It's easy to see why. The internal angles in the corners of a regular pentagon are 108 degrees. If you want the corners of adjacent tiles all to meet at a point, you will need to fit a given number of the tiles around that point, so that their angles make up a full 360 degrees. Three pentagons together produce 108 × 3 = 324 degrees, which isn't enough, so there's a gap. Four pentagons produce 108 × 4 = 432 degrees, which is too much, so there's an overlap. You just can't win.

Suppose you offset the tiles against each other, so that the corner of one tile can lie some way along a side of its neighbour. That side, being straight, takes up 180 degrees. Then, a single pentagon placed with its corner some way along the side takes up another 108 degrees. This leaves 180 – 108 = 72 degrees to be filled, which isn't enough for another pentagon. In a similar way you can show that no regular polygon with more than six sides can tile the plane.

**There are more tiling options if the tiles don't have to be regular polygons – from simple rectangles through stars (see page 169).**

# Tiling patterns

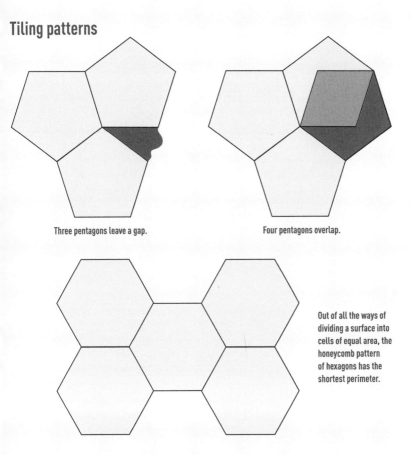

Three pentagons leave a gap.

Four pentagons overlap.

Out of all the ways of dividing a surface into cells of equal area, the honeycomb pattern of hexagons has the shortest perimeter.

While pentagons are non-tiling polygons, this isn't the case for regular hexagons, which give the familiar honeycomb pattern.

# 7.6 Aperiodic tilings

**It is harder than you think to come up with non-repeating tilings.**

Most tilings of the plane that immediately spring to mind are periodic: they consist of some basic unit that is repeated over and over in all directions.

It is possible to make tilings that aren't periodic, however. As an example, take a square tiling and divide each of the squares into two equal-sized rectangles. Use a vertical for all but one of the squares. For the last square, use a horizontal line. Whatever direction you move in, you will never encounter the horizontally divided square again, so the tiling is not periodic. This is a boring example though, because it's just the one little difference that makes the tiling **non-periodic**. The rectangles involved can equally well produce a periodic tiling.

The question is, are there sets of tiles that can *only* produce aperiodic tilings? The answer is: yes. The Penrose tiles opposite, named for the English mathematician, physicist and philosopher Roger Penrose, are an example of this. The thin and the fat rhombus can't be used to make a periodic tiling, but they do produce a non-periodic one. And unlike our boring example above, this tiling doesn't contain arbitrarily large patches that are periodic. Tilings with that property are aperiodic.

**In 1997 Penrose sued Kleenex for using a form of the Penrose tiles to emboss toilet paper – he won.**

## Sets of aperiodic tiles

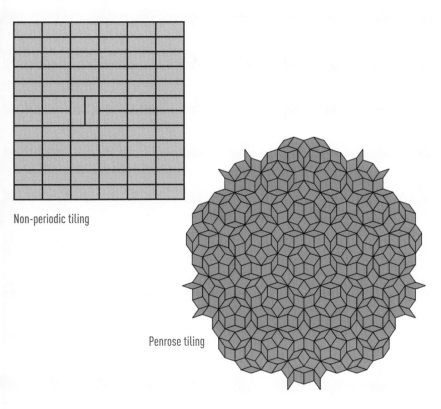

Non-periodic tiling

Penrose tiling

The top image shows a non-periodic tiling: move in any direction and you will never see the same pattern again. The Penrose tiling (bottom) is aperiodic.

# 7.7 Symmetrical solutions

**Certain problems can exhibit symmetries and being aware of this can help you find solutions.**

If you like sudoku, you'll be pleased to hear that there are 6,670,903,752,021,072,936,960 different ways of filling in a blank sudoku grid. Each of these is the solution to a puzzle, so there are plenty of games yet to play.

But wait! Given a completed grid, you can create a second one by rotating all the numbers in the first grid through 90 degrees, or by reflecting them in the vertical symmetry line of the square. Once all such symmetries are taken into account, you are left with a mere 5,472,730,538 essentially different solutions – and therefore with far fewer essentially different puzzles.

What we have discovered here is that some problems admit symmetrical solutions: once you have found one solution you may be able to create another simply by swapping around, in some way, the 'objects' that comprise the problem (numbers in the case of sudoku).

Being aware of such symmetries can be useful in a number of situations that involve solving a problem subject to constraints, such as constructing timetables or allocating resources.

**Equations can also admit symmetric solutions: both $x$ and $-x$ are solutions to $y = x^2$.**

## Symmetry in sudoku

| 1 | 2 | 5 | 3 | 7 | 8 | 9 | 4 | 7 |
|---|---|---|---|---|---|---|---|---|
| 3 | 7 | 8 | 9 | 6 | 4 | 2 | 1 | 5 |
| 4 | 9 | 6 | 1 | 2 | 5 | 8 | 3 | 7 |
| 2 | 6 | 9 | 4 | 5 | 3 | 1 | 7 | 8 |
| 8 | 4 | 1 | 7 | 9 | 2 | 6 | 5 | 3 |
| 5 | 3 | 7 | 8 | 1 | 6 | 4 | 9 | 2 |
| 9 | 1 | 2 | 5 | 8 | 7 | 3 | 6 | 4 |
| 6 | 5 | 3 | 2 | 4 | 9 | 7 | 8 | 1 |
| 7 | 8 | 4 | 6 | 3 | 1 | 5 | 2 | 9 |

| 7 | 6 | 9 | 5 | 8 | 2 | 4 | 3 | 1 |
|---|---|---|---|---|---|---|---|---|
| 8 | 5 | 1 | 3 | 4 | 6 | 9 | 7 | 2 |
| 4 | 3 | 2 | 7 | 1 | 9 | 6 | 8 | 5 |
| 6 | 2 | 5 | 8 | 7 | 4 | 1 | 9 | 3 |
| 3 | 4 | 8 | 1 | 9 | 5 | 2 | 6 | 7 |
| 1 | 9 | 7 | 6 | 2 | 3 | 5 | 4 | 8 |
| 5 | 7 | 3 | 4 | 6 | 1 | 8 | 2 | 9 |
| 2 | 8 | 6 | 9 | 5 | 7 | 3 | 1 | 4 |
| 9 | 1 | 4 | 2 | 3 | 8 | 7 | 5 | 7 |

In this example, the second grid is the first grid rotated clockwise by 90 degrees.

When programming a computer to find all sudoku solution grids, taking account of symmetry prevents the program wasting valuable search time on finding solutions that are essentially the same.

# 7.8 Noether's theorem

You might know that, in physics, certain quantities are always conserved. Noether's theorem tells us that these conservation laws are intimately related to symmetry.

**Albert Einstein called Emmy Noether a 'creative mathematical genius'.**

An example of a conserved quantity is energy. When a billiard ball strikes another billiard ball and then comes to a dead halt, its energy hasn't been lost. Rather, the first ball's energy has been transferred to the ball it struck. Other examples of conserved quantities are **momentum** (a measure of 'how much motion' there is in an object travelling along a straight line) and **angular momentum** (the analogue for rotational motion). The momentum of the first billiard ball doesn't disappear, but is transferred to the second, and the same would occur if the balls were moving around in a circular track.

*Noether's theorem* says that these conservation laws correspond to symmetries. If a physical system (a billiard table with balls) behaves the same, however it is oriented in space – so it has rotational symmetry – then this implies its angular momentum is conserved. If a system behaves the same no matter where it is located in space, it has translational symmetry, which implies conservation of momentum. And if it behaves the same no matter when it happens in time, so having translational time symmetry, then this implies conservation of energy.

Noether's theorem shows that there is a deep connection between physics and symmetry. Modern physical theories that describe the fundamental building blocks of nature are all based around the concept of symmetry.

## Angular momentum

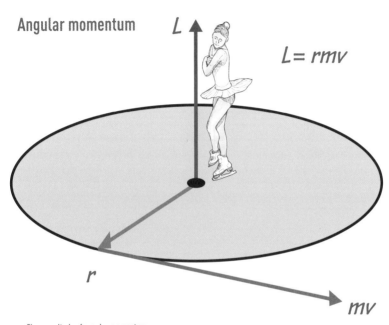

$$L = rmv$$

The magnitude of angular momentum
($L$) depends on the speed of the rotation
($v$) and the size of the object, measured by
the distance from the centre of the rotation ($r$).

Conservation of angular momentum explains why, for example a figure skater's spin speeds
up as she or he draws the arms in: as the distance ($r$) decreases the speed ($v$) must increase
so that the angular momentum ($L$) stays constant.

# 7.9 Group theory

**Collections of symmetries form interesting structures that appear throughout maths and science.**

Whenever you apply a symmetry to an object – say, a reflection to a square – and then follow it by another symmetry – say, a rotation – the result is also a symmetry. That's because leaving something unchanged twice results in it being left unchanged overall. Moreover, any symmetry can be 'undone' by another symmetry. For example, a clockwise rotation through a given angle can be undone by the corresponding counterclockwise rotation. The operation of doing nothing is (trivially) also a symmetry.

Mathematicians have taken these properties (plus a fourth one) out of their specific contexts and defined general structures called **groups**. A group is a collection of objects (which could be anything) together with an operation for combining two objects (we write + for this operation) that comply with the following rules:

1. If $a$ and $b$ are objects in the group, so is $a + b$.
2. There is an identity object $e$ so that for all other objects $a$, we have $a + e = e + a = a$ ($e$ corresponds to 'doing nothing').
3. Every object $a$ has an inverse $b$ such that $a + b = e$.
4. $(a + b) + c = a + (b + c)$ (when combining three objects in some order, it doesn't matter which of the two pairs you do first).

**The collection of whole numbers, together with the operation of addition, forms a group because it satisfies the four rules.**

# Lie groups

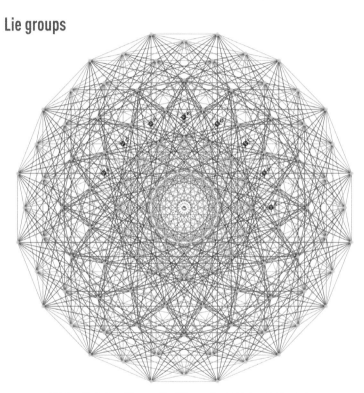

This is an illustration of the *exceptional Lie group* $E_8$, which is important in physics.

The results of group theory can be applied in whatever context
groups may appear, from investigating the symmetries of a shape
to symmetries found in the equations of physics.

# 7.10 Finite simple groups

Once mathematicians had defined groups as abstract structures, they wondered what kinds of different groups there could be. This led to the longest proof in mathematical history.

Just as the natural numbers can be expressed as products of prime numbers (see Topic 1.5), so a group can be broken up into component groups, which themselves can't be decomposed any further. These components are called **simple groups**. To understand the different structures a group can have, mathematicians in the 1970s set out to classify all finite simple groups – that is, all simple groups that are made up of finitely many objects.

The task turned out to be monumental. The classification, and the proof that it is complete and correct, spanned more than 10,000 pages, was published in around 500 journal articles and involved more than 100 different authors from around the world. Mistakes found in the original proof took another seven years to resolve and spawned further books.

The classification, finally completed in 2004, identifies 18 infinite families of finite simple groups, and another 26 individual groups, called sporadic groups, that don't fit into the 18 families. At this moment in time, only a handful of mathematicians in the world understand the whole proof. They are hard at work to come up with a simpler version so lesser mortals can grasp it, too.

**The largest sporadic group is called the Monster and is made of 808,017,424,794, 512,875,886,459,904, 961,710,757,005,754, 368,000,000,000 objects.**

## Tick Tock

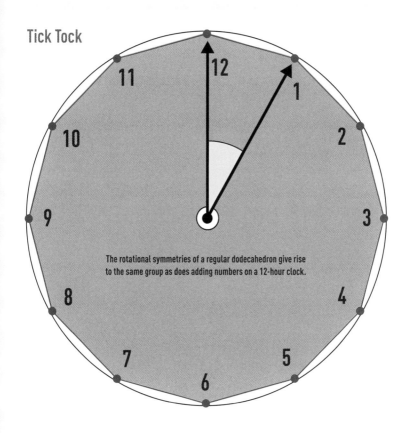

The rotational symmetries of a regular dodecahedron give rise
to the same group as does adding numbers on a 12-hour clock.

What kinds of other abstract groups are there besides the one above,
modelled on a dodecahedron? That's the kind of question addressed by
the classification of finite simple groups.

# CHANGE

**8**

$+u\frac{\partial}{\partial x}$

$=-\frac{\partial}{\partial y}+$

$t+u\frac{\partial v}{\partial x}$

$\frac{\partial w}{\partial t}+u\frac{\partial w}{\partial}$

$\frac{w}{\partial z}=-\frac{\partial P}{\partial z}+1/$

$\frac{\partial u}{\partial t}+u\frac{\partial}{\partial}$

$\frac{\partial u}{\partial z}=-\frac{\partial P}{\partial x}+1$

$\frac{\partial v}{\partial t}+u\frac{\partial}{\partial x}$

$w\frac{\partial u}{\partial z}=-\frac{\partial P}{\partial y}+$

$\frac{\partial v}{\partial t}+u\frac{\partial}{\partial x}$

$$+ \frac{\partial^2 u}{\partial y^2} + \frac{\partial^2 u}{\partial z^2}$$

$$\frac{v}{x^2} + \frac{\partial^2 v}{\partial y^2} + \frac{\partial^2}{\partial z}$$

$$\frac{\partial^2 w}{\partial x^2} + \frac{\partial^2 w}{\partial y^2} + \frac{\partial}{\partial}$$

$$\frac{\partial^2 u}{\partial x^2} + \frac{\partial^2 u}{\partial y^2} +$$

$$\left( \frac{\partial^2 v}{\partial x^2} + \frac{\partial^2 v}{\partial y^2} \right)$$

$$e \left( \frac{\partial^2 w}{\partial x^2} + \frac{\partial^2}{\partial y} \right)$$

**M**athematics gives us powerful and precise ways with which to describe change. In this chapter we start with recurrence relations, which include equations that capture the relationship between the value of something this year, say the size of a population of animals, and the next year. Such a description of the change over time is an example of a dynamical system.

These equations can display a wealth of behaviours. For example, over time the population described by such an equation may cycle through a repeating pattern of values. Or it might settle down to some individual value, known as an attractor of the system. A famous attractor, which looks like a butterfly, is the Lorenz attractor. This butterfly-shaped attractor, however, is not the source of Lorenz's famous butterfly effect. The butterfly effect was Lorenz's way of illustrating the sensitivity of

*Continues overleaf*

certain mathematical models: if you started two simulations off with slightly different starting points in your model – differing by the equivalent to just a flap of a butterfly's wing – the outcomes could be wildly different.

Such sensitivity is one of the hallmarks of mathematical chaos. Chaos provides the explanation why some dynamical systems are so difficult to describe or predict. For example, the models used to predict the weather can be chaotic. Lorenz first discovered the butterfly effect when running computer simulations of weather models. The weather is modelled using differential equations describing how changes in pressure and velocity relate to each other for any fluid. They are fiendishly difficult to solve for a given situation, and taming these equations could win you a million dollars.

# Contents

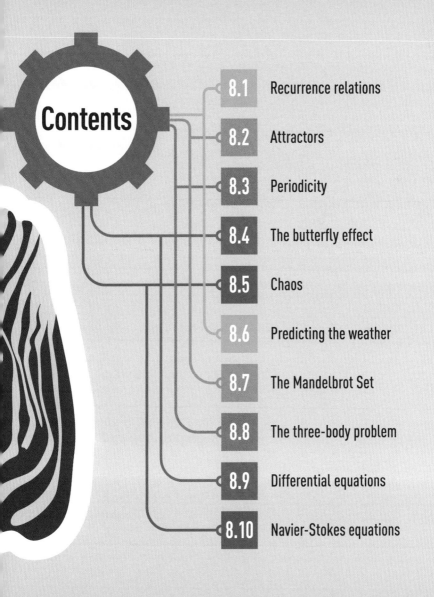

# 8.1 Recurrence relations

**How to predict the future using events from the past.**

Suppose you have a population of animals and you'd like to estimate how many there will be next year. You know that, if there are few animals, there is enough space and food for growth and the population will rise. If there are many animals, however, food will become sparse and the population will decline. A famous example of an equation describing such a situation is:

$$p_{\text{next year}} = 2p_{\text{this year}}(1 - p_{\text{this year}}).$$

Here, $p_{\text{next year}}$ and $p_{\text{this year}}$ are the proportions of animals alive (this year and next, respectively) out of some maximum number possible. This equation is an example of a **recurrence relation**, where the quantity measured at each step depends on the quantity that came before, and it displays the behaviour alluded to above. If $p_{\text{this year}}$ is less than half, there is room for growth, and $p_{\text{next year}}$ will be larger. If $p_{\text{this year}}$ is greater than half, then there are too many animals and $p_{\text{next year}}$ will be smaller. If $p_{\text{this year}}$ is exactly half, the population size stays fixed.

**Biologist Robert May first used equations like this to model population growth in the 1970s.**

This simplified model doesn't by any means describe all animal populations. Yet it is an interesting example of a **deterministic dynamical system**. Starting with any initial population you can, in theory, calculate the population sizes for all years in the future, again and again.

The graph is that of the equation in the example opposite: $y = 2x(1-x)$.

How many zebras will there be next year? The equation in our example is that of a logistic map. Generally, logistic maps have equations of the form $y = rx(1 - x)$, where $r$ is some real number.

# 8.2 Attractors

**The population model in Topic 8.1 has an interesting feature: the population size will always settle down to the same value after a few years.**

Whatever initial population proportion you start with (excluding 0 and 1), after just a few applications of the equation in Topic 8.1, the population proportion will be close to ½, and will get ever closer as you continue to apply the equation. Here, '½' is an **attractor** of the system.

Many dynamical systems have attractors. The picture on the right shows the **Lorenz attractor**. It arises from a set of equations that move points through three-dimensional space, creating complex trajectories. The Lorenz attractor is famous for its delicate, butterfly-like shape. In fact, the attractor has the complex structure of a **fractal** (see Topic 10.3). The mathematician and meteorologist Edward Lorenz developed the system in the 1960s in order to understand the behaviour of the Earth's atmosphere.

It's useful to find attractors within a dynamical system, because its presence means that you can gauge the system's long-term behaviour (for example, what size a population will eventually settle down to) – at least, for those initial values that are attracted to it.

**Edward Lorenz popularized the term 'butterfly effect' to describe mathematical chaos (see Topic 8.4).**

## The Lorenz attractor

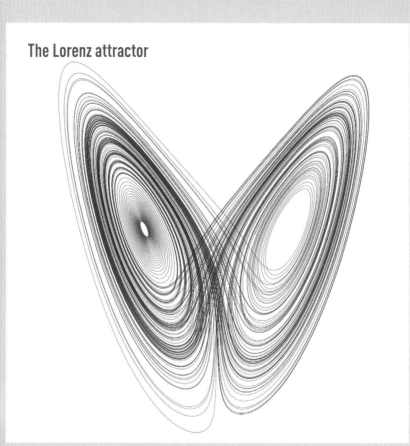

Attractors that have a fractal structure – like the one above – are called strange attractors.

# 8.3 Periodicity

**Even complicated dynamical systems can fall into regular patterns.**

As amateur billiard players will know, the path taken by a ball on a billiard table is generally pretty complicated. However, if you shoot the ball at one of the sides of the table so that it meets it at a right angle (avoiding the pocket) you can create a very predictable trajectory: the ball will move back and forth between two opposing points, retracing its path again and again. In an idealized situation, with no mass or friction impeding the ball's course, it will keep on going like this for ever.

This is an example of **periodicity**: a dynamical system falls into a regular pattern and stays trapped in it forever. Periodic behaviour occurs in many dynamical systems. Animal populations can oscillate between two or more values, the planets will retrace the same paths around the Sun for ever and ever, and a sine wave retraces the same undulating shape again and again.

Periodic behaviour can be *stable* – knock the system off its periodic path by a slight amount, and it will soon settle down again. But periodic behaviour can also be unstable, like a pencil balancing on its tip, where even the smallest change will shift it off course.

**The waxing and waning of the Moon is an example of periodic behaviour.**

# Billiard-ball trajectory

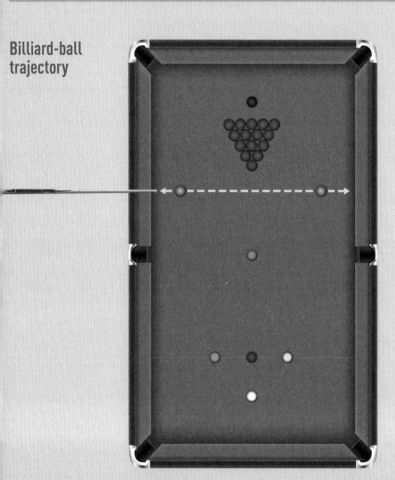

The predictable trajectory of a billiard ball that has been shot at one of the sides of a billiard table.

# 8.4 The butterfly effect

**Will the flap of a butterfly's wing in Brazil set off a tornado in Texas?**

This is the question Edward Lorenz (1917–2008) asked in his famous 1972 paper, introducing the idea of **the butterfly effect**. His reasoning was that 'two weather situations differing by as little as the immediate influence of a single butterfly will generally, after sufficient time, evolve into two situations differing by as much as the presence of a tornado'.

Mathematically, this is known as **sensitive dependence to initial conditions**. The idea is that, if you apply your mathematical equations modelling a given situation to two sets of starting values that differ only slightly, you can end up with wildly different answers. Lorenz first encountered this in 1961 when he reran a computer simulation for forecasting the weather. The first time the simulation ran, it started with an initial value of 0.506127. The second time Lorenz typed in the number by hand, rounding it down to 0.506. No one at the time would have thought such a slight difference would have an impact. But Lorenz was surprised to discover it led to wildly different forecasts.

**Lorenz initially preferred the image of a seagull causing a storm, but changed to a butterfly after Philip Merilees suggested the title for the 1972 paper.**

This sensitivity to initial conditions is now one of the hallmarks of mathematical chaos. And it doesn't only appear in complex equations such as those used to model the weather. Even simple logistic maps of the kind introduced in Topic 8.1 can display the butterfly effect.

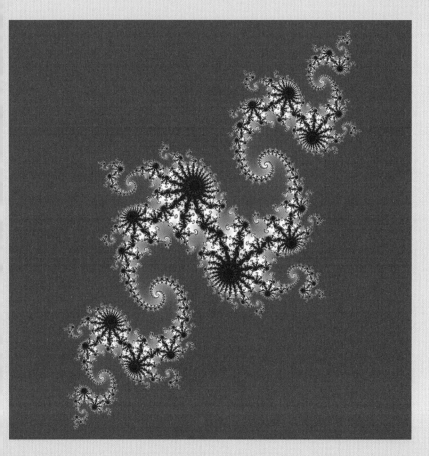

The image shows a *Julia set*, which arises from a dynamical system
closely related to the family of logistic maps. The dynamics exhibit
the butterfly effect on the boundary of the complex shape.

# 8.5 Chaos

Mathematical chaos is when 'the present determines the future, but the approximate present does not approximately determine the future'. (Edward Lorenz)

When mathematicians talk about chaos in a dynamical system, they usually mean that the system displays sensitive dependence on initial conditions (see Topic 8.4). That is, a tiny change in initial conditions (for example, today's weather) can lead to wildly different results at a later stage (next month's weather).

To understand why chaos is interesting, notice that it can occur in completely **deterministic systems** – that is, systems that don't involve any chance at all. For example, if you're familiar with the laws of physics, you should be able to exactly calculate the trajectory of a billiard ball on a billiard table (see page 187). In the real world, however, the ball isn't just an ideal point as imagined in your physics lessons. In practice, you will never be able to pin down the initial conditions – the ball's location and the force with which you hit it – with 100% accuracy. Since the smallest inaccuracy can snowball out of all proportion, this means that your calculated trajectory and the real trajectory may eventually diverge dramatically. Even a deterministic system can be unpredictable – even perfect order can give rise to chaos.

Mathematical chaos is one reason why we find it so hard to predict many real-life phenomena, like the weather or the stock market.

Mathematicians only began to fully appreciate chaos in the 1960s with the advent of computers.

Double pendulums provide a striking physical illustration of chaotic behaviour, captured in this image by Michael G. Devereux, by attaching LEDs to each part of the pendulum.

# 8.6 Predicting the weather

**Knowing if you will need an umbrella tomorrow requires grappling with chaos.**

Our weather is the result of complicated interactions of the atmosphere, the oceans and energy coming from the Sun. Weather forecasts are produced by complex mathematical models that describe these interactions using the laws of thermodynamics and the Navier-Stokes equations (see Topic 8.10). But these weather models are chaotic – a small difference in the starting conditions can result in very different forecasts (see Topic 8.4). The problem is even more pronounced, because it is impossible to measure starting conditions accurately initially to then feed them into the model to start off the forecast.

Instead, forecasters start with measurements of parameters that include temperature, wind and air pressure taken at points on a three-dimensional grid over the surface of the world. These are used as starting points for the model that then simulates the weather for a certain amount of time into the future. In order to overcome sensitivity to the initial conditions, the model simulates the weather many times, each time starting with slightly different starting conditions. This ensemble forecast predicts the likelihood of different forecasts: if 30% of the simulations forecast rain tomorrow, then the ensemble forecast is a 30% chance of rain.

**The supercomputers used for weather prediction can perform over 16,000 trillion calculations per second.**

## Charting the weather

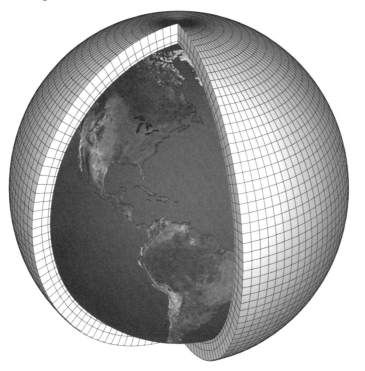

Models used for weather forecasting are based on a three-dimensional grid over the planet. The spacing of the grid is much smaller over important and densely populated sites, and much larger elsewhere.

# 8.7 The Mandelbrot set

The Mandelbrot set is an amazing structure. No matter how closely you zoom in on it, its outline remains just as crinkly as it appeared before.

The Mandelbrot set is an example of a *fractal* (see Topic 10.9), and has deep mathematical meaning. Each point $p$ in the picture opposite represents an equation $D(p)$, which describes points moving around on the plane. If $p$ lies within the Mandelbrot set, then the trajectory of the point $(0,0)$ under iterations of $D(p)$ is confined to some finite region of the plane. If $p$ is not part of the Mandelbrot set, then the trajectory of the point $(0,0)$ under iterations of $D(p)$ escapes to infinity. This simple dichotomy is what defines the infinitely complex Mandelbrot set.

The dynamical systems $D(p)$ that lie in the interior of the main black region of the Mandelbrot set all have a fixed point that is also an attractor (see Topics 8.2 and 8.3). Dynamical systems that lie in the interiors of black regions attached to the main part have periodic attractors oscillating between more than one value. Mathematicians believe that this is true for all black regions that make up the Mandelbrot set, not just the ones attached to the main one. But nobody has been able to prove that yet. This remains a very important open question in the theory of dynamical systems.

The Mandelbrot set is named for Benoit Mandelbrot (1924–2010), who discovered this, and many other fractals, in the 1970s.

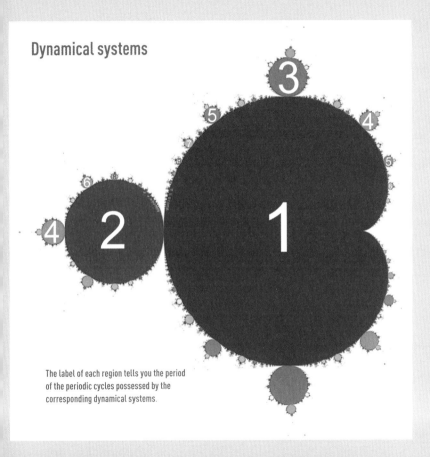

## Dynamical systems

The label of each region tells you the period of the periodic cycles possessed by the corresponding dynamical systems.

Each point in the above picture represents a complex number $p$, which in turn defines a dynamical system $D(p)$ (as explained in the text opposite). The Mandelbrot set is defined by the behaviour of the dynamical systems $D(p)$.

# 8.8 The three-body problem

**When Isaac Newton described gravity in the 17th century he also unwittingly uncovered the first example of chaos.**

It is possible to solve Newton's equations for two large bodies (such as the Earth moving around the Sun), in order to give an exact description of the orbits of these two bodies. But introduce a third into the mix and you get trouble.

The gravitational influence between three large bodies creates complications and no general formula can provide an exact description of these orbits for all time. In fact, such systems of three bodies can exhibit chaos.

If the third body is small in comparison to the other two, however, so that its gravitational influence on them is negligible, you can describe the orbits of the three bodies for some special cases in which the smaller, third body stays in a fixed position relative to the larger two bodies.

**Newton said that an exact solution for the motion of three bodies 'exceeds the force of any human mind'.**

The first three special cases were discovered by the famous 18th-century mathematician Leonhard Euler, and they describe the orbits for a smaller body that lies at three points on a straight line that passes through the two larger bodies. At roughly the same time, these three were also discovered by Joseph-Louis Lagrange (1736–1813), who in addition discovered two more points that put the smaller body on the third corner of an equilateral triangle with the two larger bodies.

## Lagrange points

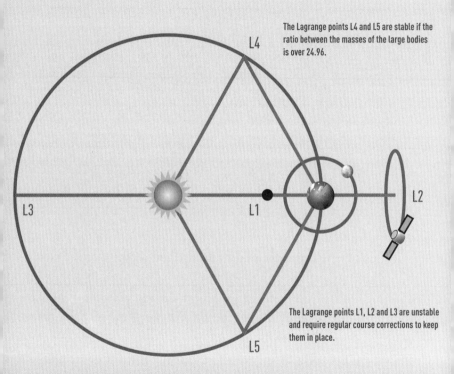

The Lagrange points L4 and L5 are stable if the ratio between the masses of the large bodies is over 24.96.

The Lagrange points L1, L2 and L3 are unstable and require regular course corrections to keep them in place.

Astronomers have used Lagrange points for their space observatories. The SOHO satellite observes the Sun at L1, and the Planck satellite observes deep space from L2.

# 8.9 Differential equations

Scientific descriptions of the world around us invariably describe how things we are interested in – speed, direction, power, energy – change.

If you drop a ball from the top of a building, how fast will it be going after ten seconds? In order to answer this question, you need to solve a **differential equation** – something that describes how the change in a value you are interested in varies with respect to a given variable. In this case, the differential equation describes how the speed of the ball changes with respect to time.

Ignoring complications such as wind resistance, we know the rate of change of the speed of any falling object is just its acceleration due to gravity: 9.8 m per second per second. So our differential equation is:

$$v'(t) = 9.8,$$

where $v'(t)$ is the rate of change with respect to time of the object's speed.

The solution to a differential equation isn't a number, it's a function – in this case, the function $v(t)$ describing the speed of the object at any time $t$. For simple differential equations, such as this one, there are standard methods for finding a solution. (The solution to our example is $v(t) = 9.8t$, so after $t = 10$ seconds the ball is travelling at 98 m per second.) But there are no known methods to solve many more complicated differential equations.

Newton's study of the gravitational orbits of three or more bodies (see Topic 8.8) was the first use of differential equations.

$$mv'(t) = 9.8\,m - kv(t)$$

A more realistic model of a falling object would be a differential equation that also includes air resistance (the constant $k$ in the equation) and mass (the constant $m$) of the object. This model would account for the fact that not all objects fall to the ground at the same speed.

# 8.10 Navier-Stokes equations

**You can observe chaos in action in the flow of water in a turbulent river.**

Drop two sticks close by each other from the river's edge and they could end up following wildly different paths downstream. The flow of any fluid – turbulent or calm – is governed by a set of equations known as the **Navier-Stokes equations**. These are differential equations (see Topic 8.9) describing how the changes in velocity (measured in three directions), pressure and viscosity of the fluid relate to each other.

The Navier-Stokes equations are more complicated than the example in the previous topic. A differential equation that only depends on one variable (such as time) is called an **ordinary differential equation**. In the Navier-Stokes equations the velocity, pressure and viscosity change with respect to time, but also with respect to position in space. Differential equations that depend on more than one variable are called **partial differential equations**.

A solution to these equations would mean you could write down an equation for the velocity and pressure for each point in space and time. As with many complicated differential equations, nobody knows of a formula that gives a solution to the Navier-Stokes equations in their most general form. Moreover, we don't even know if there is a solution, particularly one that makes sense in terms of the physical reality of fluids these equations describe.

**Finding answers to these questions about the Navier-Stokes equations would win you $1 million.**

$$\frac{\partial u}{\partial t} + u\frac{\partial u}{\partial x} + v\frac{\partial u}{\partial y} + w\frac{\partial u}{\partial z} = -\frac{\partial P}{\partial x} + 1/Re\left(\frac{\partial^2 u}{\partial x^2} + \frac{\partial^2 u}{\partial y^2} + \frac{\partial^2 u}{\partial z^2}\right)$$

$$\frac{\partial v}{\partial t} + u\frac{\partial v}{\partial x} + v\frac{\partial v}{\partial y} + w\frac{\partial u}{\partial z} = -\frac{\partial P}{\partial y} + 1/Re\left(\frac{\partial^2 v}{\partial x^2} + \frac{\partial^2 v}{\partial y^2} + \frac{\partial^2 v}{\partial z^2}\right)$$

$$\frac{\partial w}{\partial t} + u\frac{\partial w}{\partial x} + v\frac{\partial w}{\partial y} + w\frac{\partial w}{\partial z} = -\frac{\partial P}{\partial z} + 1/Re\left(\frac{\partial^2 w}{\partial x^2} + \frac{\partial^2 w}{\partial y^2} + \frac{\partial^2 w}{\partial z^2}\right)$$

$$\frac{\partial u}{\partial x} + \frac{\partial v}{\partial y} + \frac{\partial w}{\partial z} = 0$$

The Navier-Stokes equations describe how changes in velocity
(measured in three directions $u$, $v$ and $w$) relate to changes in pressure
($P$) and the viscosity of the liquid (the parameter $Re$).

# LOGIC

**M**athematics is perhaps the only field in which it is possible to be absolutely certain about something. To be accepted as valid, a mathematical result needs to be proved rigorously from a set of basic axioms, using the rules of logic. Once a result has been proven, you know for certain that it is true, always and everywhere.

In this chapter we explore the concept of mathematical truth. We will look at the rules of logic, and the idea that mathematics should be based on a tight collection of axioms: self-evident truths that nobody would call into doubt and that need no proof. We will find out how, in a world where everything is either true or false, you can assess the truth of complex statements using truth tables, and how the underlying true/false logic drives modern computers.

*Continues overleaf*

We will also take a look at two of the most important methods for proving things: proof by contradiction, which assumes that a premise that leads to a contradiction must be false; and proof by induction, which enables you to prove something about infinitely many statements at the same time, without going through them one by one. We will also find out how mathematical proofs – for millennia the prerogative of the human brain – are these days being executed by computers.

And just as you think maths has it all wrapped up, we'll introduce a dramatic result that implies that, even in maths, things aren't necessarily true or false. In some sense, maths is a matter of opinion.

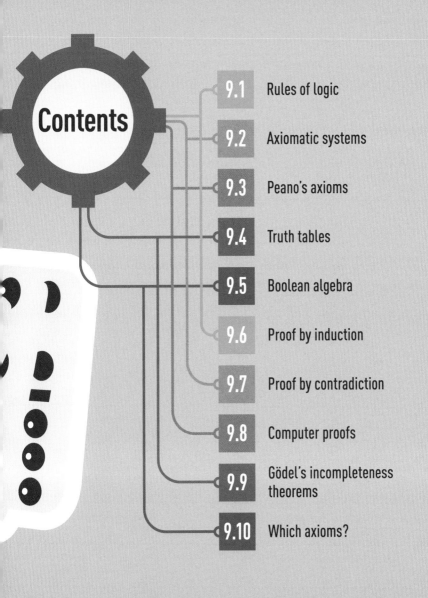

# Contents

# 9.1 Rules of logic

**We all know that the truth can be hard to come by, but are there statements that are always, and most definitely, true?**

Philosophers have pondered this question for millennia, and have come up with three ideas known as the traditional laws of thought.

The first is the law of identity, and can be phrased as saying that 'each thing is the same as itself'.

The second is the law of noncontradiction, which we can phrase as 'nothing can both be and not be'. Something is either a cat or not a cat; you either want a cup of coffee or you don't: this law seems indisputably true for most things we encounter.

The third law of thought is the law of the excluded middle: 'everything must either be or not be'. While the second law says that nothing can straddle the realms of being and non-being, the third law states that nothing can escape those two realms either.

**The three laws go back to the ancient Greeks, but that doesn't stop philosophers arguing about them today.**

These laws can be translated into **rules of logic**. For example, writing A for a statement (such as 'the Earth is round'), and NOT A for its negation ('the Earth is not round'), the first law implies that A = A. The second law implies that the two statements A and NOT A cannot be true simultaneously. And the third says that one of the statements A or NOT A must be true.

This part of the painting *The School of Athens* by Raphael depicts Plato (left) and Aristotle (right) who both played an important role in the development of the laws of thought.

# 9.2 Axiomatic systems

Ideally, mathematics should boil down to a number of axioms – self-evident truths that are used as a starting point from which all other mathematical facts can logically be deduced.

Euclid's axioms for geometry of flat space (see Topic 2.6) set out all the fundamental constructions that are possible in flat space using a compass and straightedge. All of Euclidean geometry can then be reduced to constructions based on these fundamental **axioms**.

But as we saw, Euclid's axioms are not so much fundamental truths as fundamental constraints: the fifth axiom can be varied to produce new geometries. Spherical and hyperbolic geometry are built on the same first four axioms, but the fifth axiom is different.

This illustrates the qualities of a useful **axiomatic system**. You can accept the axioms without proof or demonstration. No axioms can be derived from any combination of the others. They must be consistent and not contradict each other. They must also be the smallest set possible, yet still generate a mathematically interesting system.

Such an axiomatic system is necessary, because it allows you to use the rules of logic to prove results that you know will be true for objects in the system. It rules out hidden assumptions and holes in your arguments.

**Axioms also reveal key properties of objects in your system.**

# Axioms of origami

Mathematicians have come up with seven axioms to describe the possible constructions that can be made using straight folds in origami. These are even more powerful than Euclid's axioms.

# 9.3 Peano's axioms

**To see how axiomatic systems work, consider the basic example of the natural numbers and their arithmetic.**

Suppose you are an alien with no concept of natural numbers. Someone now gives you the following four rules:

- 0 is a natural number.
- Every natural number has a successor.
- No natural number has 0 as its successor.
- Distinct natural numbers have distinct successors.

These four axioms define all the natural numbers: 1 is the successor of 0; 2 is the successor of 1 and so on. They also define the operation of addition: 7 + 2 means 'take the number 7 and move up two steps in the list of successors'. From addition you get multiplication (repeated addition). You can then define subtraction and division (as far as possible within the natural numbers) as addition and multiplication backwards. Thus, these axioms have the power to define the natural numbers and their arithmetic.

The axioms were formulated by the Italian mathematician Giuseppe Peano (1858–1932) in 1889. He also included a fifth: *Suppose that a property holds for 0, and suppose you can prove that if this property holds for another natural number, then it also holds for the successor of that number. Then the property holds for all natural numbers.* This axiom enables you to prove things about all the natural numbers, even though there are infinitely many of them, using proofs by induction (see Topic 9.6)

**Many facts about the natural numbers and their arithmetic can be derived from Peano's axioms.**

Peano's axioms enable aliens with no concept of the natural numbers
to do sums and prove things about them.

# 9.4 Truth tables

**The three simple words NOT, OR and AND are enough to give you a whole system of logic.**

If a statement, A, is true (for example, 'I want a cup of coffee'), then the negation of this statement, NOT A ('I don't want a cup of coffee'), is false. Otherwise, if A is false, then NOT A is true. We can summarize the effect of the NOT operation in a **truth table** (see Table 1, opposite)

You can also combine two logical statements to make a third, composite statement. If you know a statement A is true (such as your desire for coffee above), then you know the statement A OR B is true no matter what the statement B says. For example, the statement '(I want coffee) OR (I want cake)' is always true if you want coffee. The only way an OR statement can be false, is if both the component statements A and B are false (see Table 2).

The other composite statement uses the AND operator. For an AND statement to be true, both the component statements A and B must be TRUE (see Table 3).

And with these three simple operations – NOT, AND and OR – you can build many more complicated statements and find out if they are true or false using truth tables.

**Modern computing is based on this binary logic, in which every statement is either true or false.**

### Cake and coffee truth tables

**Table 1: NOT**

| A | NOT A |
|---|-------|
| TRUE | FALSE |
| FALSE | TRUE |

**Table 2: OR**

| A | B | A OR B |
|---|---|--------|
| TRUE | TRUE | TRUE |
| TRUE | FALSE | TRUE |
| FALSE | TRUE | TRUE |
| FALSE | FALSE | FALSE |

**Table 3: AND**

| A | B | A AND B |
|---|---|---------|
| TRUE | TRUE | TRUE |
| TRUE | FALSE | FALSE |
| FALSE | TRUE | FALSE |
| FALSE | FALSE | FALSE |

Truth tables can reveal when composite statements, such as I want coffee (A) OR I want cake (B), are true or false. Thus making morning coffee easier to organize.

# 9.5 Boolean algebra

Two mathematicians, separated by centuries, created the language that underpins computers and our digital world.

In 1854, George Boole (1815–64) made the brilliant leap from writing logic using statements and operators such as AND, OR and NOT (see Topic 9.4) to a new type of algebra. The variables in this **Boolean algebra** have the value of 0 if they are false, and 1 if they are true:

OR is written using a new type of addition:
the rules are $0 + 0 = 0$, $0 + 1 = 1 + 0 = 1 + 1 = 1$.

AND is written using a type of multiplication:
$0 \times 0 = 1 \times 0 = 0 \times 1 = 0$, $1 \times 1 = 1$.

NOT switches the value of the variable: if $P = 1$, then NOT P, written P', is equal to 0, and vice versa. Rewriting these logical operations as algebra allows very complicated statements, that would be tedious to work through using multiple truth tables, to be solved in just a few lines.

Claude Shannon (1916–2001) made the next great leap in the 1930s. He realized that the complicated circuits used in telephone exchanges could be seen as a physical embodiment of Boole's algebra. The circuits could take two values: they were either closed (with a value of 1) or open (with a value of 0). The arrangements of switches operated in the same way as the OR, AND and NOT operators of addition, multiplication and negation in Boolean algebra.

Shannon realized that all information could be represented by series of 'bits', digits with a value of 0 or 1.

## Simplifying circuits

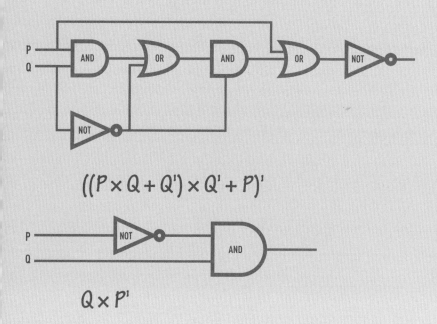

$$((P \times Q + Q') \times Q' + P)'$$

$$Q \times P'$$

You can use Boolean algebra to prove that the complicated circuit, corresponding to the expression $((P \times Q + Q') \times Q' + P)'$, can be simplified to the far simpler circuit, written as $Q \times P'$.

# 9.6 Proof by induction

How do you prove something is true for all the natural numbers, even though there are infinitely many of them?

Think of a **proof by induction** as a line of dominoes all poised to topple. Suppose you want to show that the sum of the natural numbers from 1 to $n$ is equal to $\frac{n(n+1)}{2}$. We can see it is true for the first domino: when $n = 1$ the sum is just 1, and the value of:

$$\frac{n(n+1)}{2} = 1 \times \frac{(1+1)}{2} = 1.$$

Once you have established the truth of the statement at the beginning of the natural numbers, you then need to line up your dominoes. One domino can knock over the next, if the statement is true for some number $n$ also implies that the statement is true for $n + 1$. For our example, assuming the statement is true for $n$ gives:

$$(1 + \ldots + n) + (n + 1) = \frac{(n(n+1))}{2} + (n + 1).$$

We can simply rearrange this to show the statement is true for $n + 1$:

$$1 + \ldots + n + n + 1 = \frac{(n+1)(n+2)}{2}$$

If your statement being true for the $n$th domino means it's also true for the $(n + 1)$th domino, then you know that starting the first domino falling will cause each one to knock over its successor, showing it's true for all the natural numbers.

The ancient Greek scholar Plato was among the first to use proof by induction, in his work *Parmenides* in 370 BC.

Proof by induction is similar to a line of dominoes arranged so that each one will knock over the next.

# 9.7 Proof by contradiction

If a given assumption implies a contradiction, then surely the assumption itself must be false.

Consider the proof that there are infinitely many prime numbers. Start by assuming that the opposite is true: that there are finitely many prime numbers. Then we can list them by order and give them the names $p_1$, $p_2$, $p_3$ and so on, all the way up to the (supposedly) largest prime $p_n$. Now consider the number $E$ you get by multiplying all those primes and adding 1:

$$E = p_1 \times p_2 \times p_3 \times \ldots \times p_n + 1.$$

The number $E$ is bigger than all the prime numbers in our list, and since the list includes all the primes there are, this means that $E$ can't itself be a prime number. However, as we know from the fundamental theorem of arithmetic (see Topic 1.5), $E$ can be written as a **product** of prime numbers. This means that each prime factor of $E$ would also have to be one of the primes on our list, because the list includes all primes there are. However, from the formula above, we see that dividing $E$ by any of the primes on the list leaves a remainder of 1. Therefore, our list doesn't contain all the primes. This is a contradiction.

Our assumption that there are only finitely many primes must therefore be false. It follows that there are infinitely many prime numbers.

The ancient Greek mathematician Euclid formulated this proof around 300 BC.

## The excluded middle

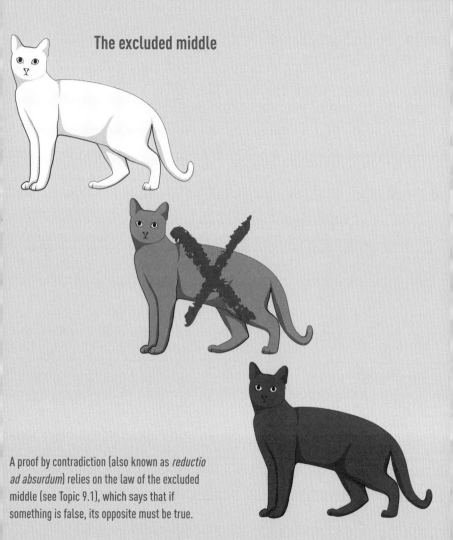

A proof by contradiction (also known as *reductio ad absurdum*) relies on the law of the excluded middle (see Topic 9.1), which says that if something is false, its opposite must be true.

# 9.8 Computer proofs

**Is a mathematical proof still valid if it lies in the brain of a computer rather than the brain of a human being?**

For millennia, mathematical proof has been thought of as something that can be derived from a series of logical steps from the axioms that define the system being examined. This can produce proofs of massive length (such as the **classification of finite simple groups** from Topic 7.10 and Andrew Wiles' proof of Fermat's last theorem from Topic 3.10), but a person could, theoretically, understand the proof from start to finish.

A challenge to this concept of proof arose in 1976 when Kenneth Appel (1932–2013) and Wolfgang Haken (b. 1928) proved the four-colour theorem (that any flat map can be rendered in just four colours so that no adjacent countries are the same colour). The traditional part of their proof reduced their problem to a large number of special cases, but then relied on the brute force of a computer to check each of these. Many mathematicians at the time claimed it wasn't a proof at all. Although they could check each step of the computer's program, no human could possibly check each of the computer's calculations – otherwise, they would have not needed the computer in the first place.

Mathematicians can no longer ignore computer proofs, since the proofs of other important mathematical results have been discovered that also rely on computers.

**Computer proofs are now treated in a similar way to experiments – they are accepted if they are verified and their results replicated.**

In 1998, Thomas Hales used a computer to prove Kepler's 1609 conjecture that the greengrocer's method of stacking oranges is the most efficient.

# 9.9 Gödel's incompleteness theorems

**In maths everything is either true or false . . . or is it?**

Throughout the ages, maths has always been a jumble of different fields: geometry, algebra, calculus and so on. So at the beginning of the 20th century mathematicians decided a clean-up was necessary. Their aim was to base all of mathematics on a definitive set of axioms and show that there are no contradictions.

In the 1930s, however, the Austrian mathematician Kurt Gödel (1906–78) delivered a bombshell that dashed this axiomatic dream. Suppose you have such a formal axiomatic system that contains the natural numbers and their arithmetic. Then, so Gödel proved, there will always be statements about those numbers that you can **formulate** within your system, but that you can't prove to be either true or false from its axioms.

When you come across such an **undecidable statement** within a formal system and you think it should be true, you can of course decree it to be true and add the decree to your axioms. However, **Gödel's incompleteness theorems** implies that this will either cause a contradiction, or other undecidable statements will remain. No matter how cleverly you choose your axioms, there will always be undecidable statements. Goodbye to the beautiful notion of definite mathematical truth.

**The undecidable statements mathematicians have found don't affect day-to-day maths.**

Kurt Gödel was a good friend of Albert Einstein. Later in life he worked on Einstein's theory of relativity and showed that, in theory, there could be universes in which time travel into the past is possible.

# 9.10 Which axioms?

**If every set of axioms produces undecidable statements, then on what set of axioms should mathematics be based?**

It's a good question. At the start of the 20th century mathematicians, notably the British polymath Bertrand Russell (1872–1970), realized that all mathematical objects can be described as collections of 'things' – such collections are called sets. They tried, therefore, to phrase the axioms of mathematics in terms of **set theory**. It was a monumental effort that eventually culminated in Russell's three-volume work *Principia Mathematica*, written with Alfred North Whitehead (1861–1947) and published between 1910 and 1913. The machinery developed in *Principia* is so cumbersome that 1 + 1 = 2 can't be proved until well into the second volume. Russell and Whitehead commented on this particular result that 'the above proposition is occasionally useful'.

Russell and Whitehead's system was eventually superseded by one based on the so-called ZFC axioms, named for Ernst Zermelo (1871–1953) and Abraham Fraenkel (1891–1965), and a special rule called the **axiom of choice**. Mathematicians who are bothered with the foundations of maths seem broadly to agree that the ZFC axioms are the way forwards. Of course, because of Gödel's result (see Topic 9.9), there are undecidable statements within the ZFC system, but mathematicians are busy looking for additional axioms to settle at least those that they find most pressing.

> **Most working mathematicians don't worry about these foundational problems, or perhaps only on Sundays.**

Undecidable statements cannot be avoided. It's like a giant jigsaw puzzle that won't ever fit together properly. Ultimately you have to develop an axiomatic system of your own.

# INFINITY

**A**t first, infinity may seem to be incomprehensible, but that has never put off mathematicians. They have been exploring infinity for millennia, finding concrete ways to get to grips with this most elusive mathematical concept.

There are, in fact, many different types of infinity. Philosophers distinguish between a potential infinity – the destination of a process that you'll never reach, such as the largest counting number you can think of – and actual infinities that could exist in the real world. Usually a prediction of infinity in a scientific theory describing the real world indicates the point at which that theory breaks down. But there are places, such as black holes, where some scientists believe infinity may actually exist in nature. Perhaps the only place we'll ever see a physical picture of infinity is with the infinitely repeating images of fractals.

*Continues overleaf*

Mathematicians also have more concrete questions. For example, how big is infinity? In this chapter we'll find a clever way of pairing the elements of infinite sets to compare the size of one infinity to another. We'll see how this remarkably straightforward argument proves that two infinities we're very familiar with – the counting numbers and the real numbers – are not the same size.

One of the great challenges to maths, known as the continuum hypothesis, asks if there are any different sized infinities between the counting numbers and the real numbers. It's not possible to answer this question within our current framework of mathematics – we'll need a still-deeper understanding of mathematics to answer it.

# Contents

# 10.1 Potential and actual infinity

The Greek mathematician and philosopher Aristotle distinguished between two types of infinity: a potential infinity and an actual infinity.

We all have an intuitive grasp of what infinity is: it's something that is associated with things that never end. One example is a never-ending expanse of space. No matter for how long you travel, you will never reach the end of it. Another example is the never-ending sequence of natural numbers 1, 2, 3, 4, etc. You can keep on counting forever, but you will never finish. In these examples, you don't encounter infinity, rather it lurks at the end of something that never ends. This kind of infinity is called a **potential infinity**.

Each thing that makes up a potential infinity – for example, each number or each point in space, is itself finite. Aristotle phrased it beautifully in his book *Physics*:

*'For generally the infinite has this mode of existence: one thing is always being taken after another, and each thing that is taken is always finite, but always different.'*

By contrast, an **actual infinity** is one that (if it exists) you do encounter. For example, if the value of something – say, the density of matter inside a black hole – became infinite in a particular point in space and time, then that would constitute an actual infinity.

**Aristotle believed that actual infinities couldn't exist in nature.**

It's tempting to think of the universe as a potential infinity: one that extends infinitely in all directions. Physicists, however, are not sure that this is really true, or that the universe is actually finite.

# 10.2 Infinity in nature

**It's fair to say no human being has ever seen infinity, but could it actually exist?**

Scientists usually assume that a prediction of some actual value of infinity in the world is a problem with the mathematical models they are using. Perhaps the model is too simple, and the infinity will disappear if they include . more information. Or perhaps the theory no longer applies at the point infinity is predicted.

However, there are some predictions of **infinity in nature** that do seem plausible to some scientists. In cosmology, for example, theory suggests that the universe began with a Big Bang, when the universe was infinitely dense, infinitely small and infinitely hot. The theories also predict black holes. These should still exist today, and indeed there is thought to be one at the centre of almost every galaxy. Black holes are predicted to have infinite density and infinite gravitational attraction at their centre.

But what does infinity look like, if it exists? We certainly won't ever know what the infinity inside a black hole looks like: it is shielded from our observations by the horizon – the region of space surrounding the black hole that marks the point at which nothing, including light or information, can escape its gravitational pull.

**'Past time is finite, future time is infinite.' (Edward Hubble, the first person to observe the expanding universe, 1937).**

While many scientists view the predictions of infinity with a degree of scepticism, others believe that there are real, physical infinities and that they play an important role in the structure of the universe.

# 10.3 Fractals

Fractals are self-similar structures that exhibit the same complexity no matter how closely you zoom in on them. Many of them come from simple, yet infinite, processes.

Take a square and divide it up into nine smaller squares. Remove the centre square, so that you now have eight squares. Divide each of those into nine smaller squares and, again, remove the centre square from each one, to give you 8 × 8 = 64 smaller squares. Repeat the process with each of these and keep going, ad infinitum. At every step, divide each square into nine smaller ones and remove the centre square.

When you have finished (obviously, you never will, but use your imagination), you will be left with a strange shape, call it $X$, that's **self-similar**: it looks the same no matter how much you zoom in. That's because the same infinite procedure (of removing squares) is applied to any square you come across in the construction.

The shape $X$ is so full of holes that it contains no area at all. Since it covers no area, $X$ can't be said to be two-dimensional. On the other hand, $X$ is much more complex than an ordinary one-dimensional line or curve. Mathematicians have settled on a new way of defining the dimension of these kinds of strange objects – and with this new definition our shape $X$ has a dimension of 1.8928.

**A fractal is defined by the fact that its dimension is not a whole number: it is fractionally dimensional.**

## The Sierpiński carpet

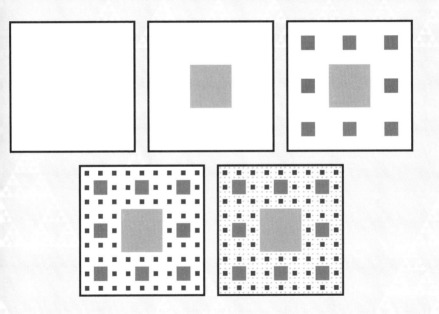

The first few steps towards the shape *X*. It is called the Sierpiński carpet, after Wacław Sierpiński, who first described it in 1916. The fractal in the background of this image is produced in a similar way, but starting with a triangle rather than a square.

# 10.4 Cardinality

It's possible to compare the 'sizes' of two infinite collections of things by seeing if you can exactly pair the objects that make them up.

Imagine you have a room full of people and chairs. If every person gets to sit on exactly one chair and there are no chairs left over, then you know that there are as many chairs as there are people.

Mathematicians apply the same idea to infinite collections of things – for example, an infinite collection of chairs and an infinite collection of people. If there is a way of pairing each object in one collection with exactly one object in the other, so that no objects are left over in either collection, then they say that the two collections have the same 'size', or **cardinality**. This makes sense, but can lead to strange results. Notice that you can pair the even numbers exactly with the natural numbers:

2 is paired with 1 because it's the 1st even number,
4 is paired with 2 because it's the 2nd even number,
6 is paired with 3 because it's the 3rd even number,

and so on. By our definition above this means that the collection of even numbers has the same cardinality as the collection of all positive natural numbers – even though you'd think there should only be half as many! It's a strange result, but mathematicians have come to accept it.

**Galileo Galileo was put off thinking about infinity by these kinds of results.**

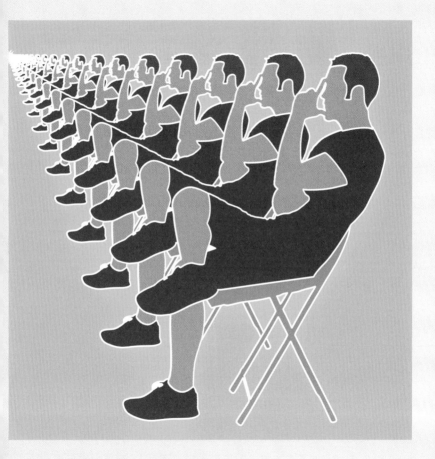

An infinite number of people on an infinite number of chairs.
The two infinite collections can be compared in terms of size
and have the same cardinality.

# 10.5 Countable infinity

**How do you count infinity? Obviously, you use the natural numbers . . .**

Did you ever try to count to infinity as a small child? While no one has ever succeeded in doing this, it isn't as foolish an enterprise as you might at first think.

The infinity of the natural numbers is the first one most of us encounter. Although a difficult concept to consider, it is at least a nice, neat sort of infinity. We have a clear map of how to get there: you start at 1, next is 2, then 3, then 4 and so on.

This ability to list the natural numbers in such a well-ordered, exhaustive fashion led the German mathematician Georg Cantor (1845–1918) to call this type of infinity a **countable infinity**. And it isn't just the natural numbers that are countably infinite. You can neatly list the even numbers (see Topic 10.4) – first is 2, then 4, then 6, then 8 and so on. You could list the odd numbers in a similar way (the $n$th number on the list is $2n - 1$). In fact, if you can list the members of an infinite set in this way, you can also pair them with the natural numbers – the first number on the list is paired with 1, the second with 2 and so on. Such a set has the same 'size', or cardinality, as the natural numbers, and so is also countably infinite.

**The countable infinity of the natural numbers is the smallest infinity.**

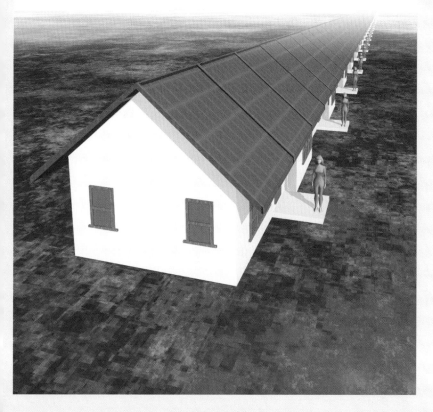

Hilbert's hotel is a thought experiment in which any finite number, *n*, of
new guests can be housed, by getting everyone to move on by *n* rooms.
It allows you to accommodate countable infinities of new guests.

# 10.6 Rational numbers and infinity

Sets that, at first, seem smaller, or larger, than the natural numbers, can actually also be countably infinite.

One of the strange things about countable infinity is that sets that seem obviously smaller than the natural numbers are actually the same 'size', or cardinality. For example, the order in which you list the even numbers immediately gives you a pairing with the natural numbers (see Topic 10.4).

You can also write a complete list of the rational numbers – that is, fractions $p/q$ where $p$ and $q$ are natural numbers. At first consideration this might seem impossible. There are infinitely many fractions between just the natural numbers 1 and 2: 1/2, 1/4, 1/8, 1/16 and so on. But there is a clever way of counting them that shows the rational numbers are also countably infinite.

To do this, write the rational numbers in a grid, with the number in row $i$, column $j$, being the number $i/j$. This grid includes every rational number, including many repeats, such as 1/1 = 2/2 = 3/3, and so on. You can then list the rational numbers by following the wiggling diagonal line shown opposite. There are some repeats in the list, but you can drop any number from the list that has been represented before. What remains is a pairing of the natural numbers with the rationals – therefore, the rational numbers are also countable.

**The German mathematician Georg Cantor first proved this theory in the 1870s.**

## Working with rational numbers

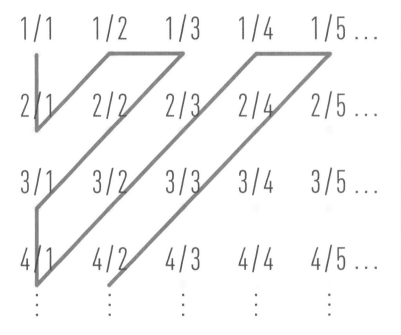

It is possible to write a well-ordered, exhaustive list of the rational numbers using this grid.

# 10.7 Uncountable infinity

**There exists an infinity that is 'bigger' than the infinity of the natural numbers. Illustrated by the real numbers, it is called an uncountable infinity.**

The real numbers are all the numbers you find on the number line, including natural numbers, negative numbers, rational numbers and irrational numbers. If you imagine an infinitely long ruler, then every point on the ruler gives you a real number and every real number gives you a point on the ruler.

The natural numbers form a collection of discrete points on our infinitely long ruler, at distance 1 apart. This suggests that the infinity of the real numbers is 'bigger' than the infinity of the natural numbers: the natural numbers have identical gaps between them, while the real numbers manage to fill those gaps. They form a **continuum**.

It turns out that this intuition is correct. You can't make a list of the real numbers that pairs them, one by one, with the natural numbers. As you will discover in the next topic, such a list is bound to miss out at least one real number. This means that the set of real numbers has a larger cardinality than the set of natural numbers. The infinity of the real numbers is called an **uncountable infinity**.

Every infinite collection that cannot be put in 1:1 correspondence with the natural numbers is called 'uncountable'.

**Even just the real numbers between 0 and 1 are uncountably infinite.**

## Infinities within infinity

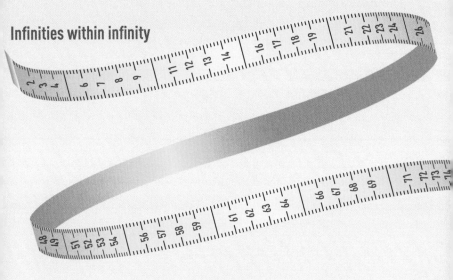

This tape also accommodates all the irrational numbers, including:

$$\sqrt{2} = 1.4142135623730 95\ldots$$
$$e = 2.718281828459045\ldots$$
$$\pi = 3.141592653589793\ldots$$

The real numbers, represented by this measuring tape, include all numbers represented by the usual marks such as 1, 2 and 3 and some numbers in between, such as 1.4, 2.8, 3.2 and so on. Hiding between these marks are the irrational numbers, including $\sqrt{2}$, $e$ and $\pi$.

# 10.8 Real numbers and infinity

**Using a proof by contradiction, it is possible to prove that the real numbers are uncountable.**

Let's start by assuming that the **real numbers** are countable – that is, they can be put in one-to-one correspondence with the natural numbers. This would mean that we can make a list of all the real numbers, for example, the list might start like this: 0.23456 . . .; 3.67896 . . .; -6.65434 . . . ; 0.8566 . . . and so on, where the dots indicate that decimal expansion may continue, possibly forever. Obviously, our list will be infinitely long. (You also need to account for ambiguities, for example, that 0.999. . . = 1, but that's easily done.)

Now make a new number by starting with a 0, followed by the decimal point, followed by the *first* digit after the decimal point of the first number, followed by the *second* digit after the decimal point of the second number and so on: 0.2746 . . . in our example. Now increase each digit by 1 (or write a 0 if the digit is 9): in our example, this gives 0.3857 . . .

**This proof is known as Cantor's diagonal argument, after the mathematician Georg Cantor.**

This number is different from the first number on the list, because it differs in the first digit after the decimal point. It's different from the second number on the list, because it differs in the second digit after the decimal point and so on. In fact, it is different from *all* numbers on our list, which means it's not on the list. This is a contradiction because we assumed our list contained *all* real numbers. Therefore, our original assumption is false and the real numbers are uncountable.

## Cantor's diagonal argument

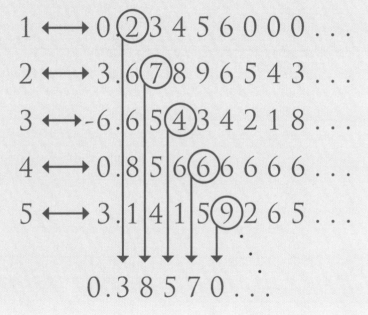

Cantor's diagonal argument presented a mathematical proof that certain numbers exist, that cannot be put into one-to-one correspondence with the infinite set of natural numbers.

# 10.9 Cantor's paradise

Georg Cantor revealed infinitely more infinities than just the countable infinity of the natural numbers and the uncountable infinity of the real numbers.

A *set* is a collection of objects. For example, the numbers 1, 2 and 3 form a set. A **subset** of a set is just another collection, possibly smaller, made of (some) of the same objects. For example, the subsets of our set $S = \{1, 2, 3\}$ are $\{1\}$, $\{2\}$, $\{3\}$, $\{12\}$, $\{13\}$, $\{23\}$, $\{123\}$ and the empty set, which has nothing in it. The collection of subsets of a set, $S$, form a set themselves – it's called the **power** set of $S$.

How big is the power set of a set? The cardinality, or 'size' (see Topic 10.4), of a power set of a set $S$ is always 2 raised to the power of the cardinality of $S$. This works for finite sets, such as our example: the cardinality of $S$ was 3, and the cardinality of the power set of $S$ was $2^3 = 8$.

This also works for infinite sets. For example, the cardinality of the power set of the natural numbers is $2^{\aleph_0}$, where $\aleph_0$ is the cardinality of the natural numbers ($\aleph$ is the Hebrew symbol, aleph). Cantor showed that you can never pair the members of a set $S$ with the members of its power set: the cardinality of the power set is always strictly 'larger'.

In this way, Cantor revealed an infinity of infinities. Starting with the set of natural numbers, and successively taking power sets, produces an infinite list of infinite sets, each with cardinality larger than the last.

**The real numbers have the same cardinality as the power set of the natural numbers.**

## Power sets

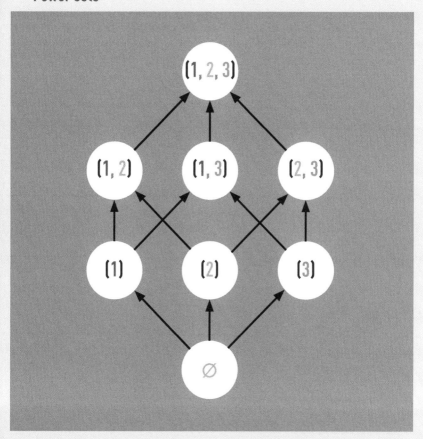

Cantor developed an arithmetic for these infinite numbers. While many mathematicians were horrified by his work, David Hilbert said: 'No one shall expel us from the paradise that Cantor has created'.

# 10.10  The Continuum Hypothesis

**We know that the cardinality of the natural numbers is strictly smaller than the cardinality of the real numbers – but does another infinity exist between these two?**

In the 1870s, Cantor revealed that there was a hierarchy of infinities. The first, the smallest, is $\aleph_0$, the cardinality of the natural numbers.

The next we know about is the cardinality of the real numbers: $2^{\aleph_0}$. What is still unknown is whether or not there are any infinities between these two.

The **continuum hypothesis** (so-called because the real numbers are also known as the 'continuum') states that there isn't: the next biggest infinity after the cardinality of the natural numbers is that of the real numbers.

A more concrete way of thinking about this is to ask whether there is a subset of the real numbers with a cardinality that is between $\aleph_0$ and $2^{\aleph_0}$.

This question remains unanswered today. Moreover, it is now known that it cannot be answered within the axioms of mathematics currently in use (see Topic 9.10). This upsets some mathematicians, who feel this is an important question and that we should be able to answer it. So the hunt goes on for a new axiom that settles the continuum hypothesis.

**Cantor's work essentially started the field of set theory, one that is still flourishing today.**

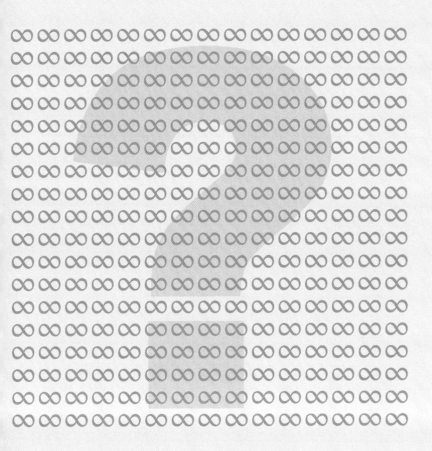

Unlike other cases of famous problems, such as Fermat's last theorem,
the Continuum Hypothesis will only be proved or disproved following
fundamental changes to the foundations of mathematics.

# Glossary

**Axiom**
A mathematical statement that can be accepted without question and used as a premise for developing mathematical arguments.

**Calculus**
The branch of maths concerned with change. Differential calculus is concerned with rates of change (expressed in the slope of a graph); integral calculus is concerned with the accumulation of quantities linked in changing situations (expressed in the area beneath a graph).

**Chaos**
The tendency of dynamical systems to be unpredictable.

**Constant**
Any number with a specific, unchanging value appearing in a mathematical equation, or with significance in maths in general.

**Convergence**
The tendency of terms in a mathematical sequence or series to approach a single limiting value.

**Coordinates**
A system of reference that allows the definition of unique points in space – for example, points in the two-dimensional plane, or in three-dimensional space.

**Divergence**
The tendency of terms in a mathematical sequence to approach infinity.

**Dynamical system**
A system that changes over time, such as the weather.

**Equation**
A balanced mathematical equality. In other words, a relationship in which the terms on either side – numbers, variables and constants – are equal to each other.

### Fractal

A shape that is (approximately) self-similar and whose dimension is not a whole number.

### General solution

A single formula that gives a solution to a whole family of equations that are all of a particular type. For example, the quadratic formula gives you the solutions to all quadratic equations.

### Geometric series

A mathematical series in which there is a common ratio between the successive terms.

### Harmonic series

A mathematical series in which the $n$th terms is $1/n$. The series is divergent, and linked to the overtones and harmonics found in music.

### Imaginary number

A number that is a multiple of the imaginary number $i$ (the square root of $-1$). The number $i$ cannot be found on the real number line.

### Irrational number

A real number that cannot be written as a fraction.

### Limit

The value a sequence or series converges to.

### Natural number

One of the counting numbers 1, 2, 3, etc.

### Partial sum

The sum of all the terms in a mathematical series up to a particular term.

### Polygon

A shape drawn on the plane bounded by a number of straight lines.

### Positional number system

A way of writing numbers in which the value of a symbol depends on its position.

### Power

A number written as a superscript after another number, which shows how many times that number should be multiplied by itself.

**Prime number**
A number that is divisible only by itself and 1.

**Product**
The result of multiplying two or more numbers together.

**Proof**
A watertight logical argument showing that a statement is true.

**Rational number**
A number that can be written as a fraction.

**Real number**
A number that represents a point on a continuum stretching from minus infinity to infinity.

**Self-similarity**
A property of objects in which the whole object is similar or identical to one of its parts, so the object displays the same properties on many scales.

**Series**
A sum with infinitely many terms.

**Set**
A collection of objects such as numbers or geometric figures. Considering the abstract properties of sets is one of the most fundamental concepts in maths.

**Subset**
A set whose members are all members of another set.

**Symmetry**
A property of objects (such as geometric figures) whereby they remain unchanged when subjected to a transformation such as rotation, reflection or translation.

**Undecidable statement**
In mathematical logic, a statement that cannot be proved or disproved from the accepted axioms in a system.

**Variable (dependence)**
An element of an equation whose value is either arbitrary or unknown, usually represented by a letter such as $x$ or $y$. Variables may be dependent or independent. Dependent variables have values that depend on the value of one or more others.

# Index

# Acknowledgements

The authors would like to thank all the people at the Millennium Mathematics Project (mmp.maths.org) and *Plus Magazine* (plus.maths.org) for enabling us to explore the wonders of mathematics for the last 15 years.

## Picture Credits

Quantum Books Limited would like to thank the following for supplying the images for inclusion in this book:

7, 51 Jos Leys – www.josleys.com; 15 The Royal Belgian Institute of Natural Sciences, Brussels; 47 Charles Trevelyan; 63 DEA PICTURE LIBRARY; 71 Shyshell; 75 The Opte Project; 81 Adam Cunningham and John Ringland via Wikipedia; 87 Shutterstock/Hadrian; 93 Shutterstock/keko-ka; 105 Wikimedia Commons; 111 Shutterstock/images72; 113 Shutterstock;Robert Adrian Hillman; 119 Shutterstock/studiostoks; 125 Shutterstock/Yuganov Konstantin; 129 Shutterstock/isak55; 135 Shutterstock/ IR Stone; 145 Eric Gaba (Sting) via Wikipedia; 149 Shutterstock/Dan Breckwoldt; 151 Adam Weyhaupt, Southern Illinois University Edwardsville; 159 Shutterstock/konmesa; 165 Shutterstock/Carlos Amarillo; 175 Wikimedia Commons; 183 Shutterstock/ Chantal de Bruijne; 185 User:Dschwen. via Wikipedia; 187 Shutterstock/romvo; 189 Ve4cib via Wikipedia; 191 Michael G. Devereux; 193 Wikimedia Commons; 195 Wikimedia Commons; 201 Shutterstock/ block23; 207 Wikimedia Commons; 209 Shutterstock/Mara008; 211 Shutterstock/ Anna Bogatireawa; 217 Shutterstock/ TijanaM; 221 Shutterstock/Alexei Novikov; 223 EMILIO SEGRE VISUAL ARCHIVES/AMERICAN INSTITUTE OF PHYSICS/SCIENCE PHOTO LIBRARY; 231 Shutterstock/Anton Jankovoy; 233 Shutterstock/Hallowedland; 237 Shutterstock/Ficus777.

While every effort has been made to credit contributors, Quantum Books Limited would like to apologize should there have been any omissions or errors and would be pleased to make the appropriate corrections to future editions of the book.